Shale Analytics

Shahab D. Mohaghegh

Shale Analytics

Data-Driven Analytics in Unconventional
Resources

 Springer

Shahab D. Mohaghegh
Petroleum and Natural Gas Engineering
West Virginia University
Morgantown, WV
USA

and

Intelligent Solutions, Inc.
Morgantown, WV
USA

ISBN 978-3-319-84008-6   ISBN 978-3-319-48753-3   (eBook)
DOI 10.1007/978-3-319-48753-3

This Springer imprint is published by Springer Nature
The registered company is Springer International Publishing AG
The registered company address is: Gewerbestrasse 11, 6330 Cham, Switzerland

To Dorna that gives my life meaning

Foreword

It is an honor and a pleasure to be asked to write the Foreword to this much anticipated book on the soft-computing, data-driven methodologies applied across unconventional reservoirs so as to harness the power of raw data and generate actionable knowledge. We are taken along a well-documented but still bumpy road that starts with an introduction to the shale revolution and draws salient comparisons between the traditional modeling of these unconventional resources and the non-deterministic and stochastic workflows prevalent in all industries that strive to analyze vast quantities of raw data to address and solve business problems.

We are enlightened as to an array of analytical methodologies that have successfully proven to be not only pertinent in the oil and gas industry but also computer resource friendly. Methodologies drawn from artificial intelligence and data mining schools of thought, such as artificial neural networks, fuzzy logic, fuzzy cluster analysis and evolutionary computing, the last of which is inspired by the Darwinian Theory of Evolution through Natural Selection.

The book inspires geoscientists entrenched in first principles and engineering concepts to think outside the box shaped by determinism, and to marry their experience and interpretations of the data with the results generated by data-driven advanced analytical workflows. The latter approach enables ALL the data to be exploited and opens the door to hidden relationships and trends often missed by the former approach executed in isolation. We are encouraged as engineers to not only to break out of the silos and put ALL our data into the context of experience and interpretation but also to allow the data to do the talking and ask questions of our traditional workflows.

The convincing conversational tone is enhanced by some in-depth case studies that have proven successful to oil and gas operators from a business value proposition perspective. The Shale Production Optimization Technology (SPOT) chapter alone is validation enough to provide credible endorsement of the strategic and tactical business insight into hydraulic fracturing practices essential to exploit the unconventional resources. The book provides plausible and cogent arguments for a top-down modeling methodology. We are introduced to a very convincing presentation of material that details a "formalized, comprehensive, multivariate, full-field, and empirical reservoir model".

We are fortunate to have this book published at such an opportune time in the oil and gas industry as it climbs slowly out of a big trough left by the erratic price fluctuations of the past two years since 2014. Operators and service companies alike will see incredible value within these pages. The authors have diligently provided an index to very important soft-computing workflows to ensure ALL the hard and soft data detailed in the early chapters are mined to surface powerful knowledge. This original knowledge can help design best practices for drilling and completing the wells in unconventional reservoirs. It also sheds light on different modeling practices for field re-engineering and well forecasting in reservoirs that necessitate innovative workflows as traditional interpretive approaches honed in the conventional reservoirs prove inadequate.

The strength of the messages logically laid out in this book introduce the reader to a collection of soft-computing techniques that address some critical business issues operators in the unconventional reservoirs are facing on a daily basis: quantifying uncertainty in the shale, forecasting production, estimating ultimate recovery, building robust models and formulating best practices to exploit the hydrocarbons.

We are indebted to Shahab Mohaghegh for his original thought, passion and innovation across the many years as he evangelizes the application of the ever-increasing popularity of data-driven methodologies. As one of the earliest pioneers in this field, we are grateful that he has put pen to paper and provided the industry with a very valuable book.

Cary, USA   Keith R. Holdaway FGS
                                                                 Advisory Industry Consultant
                                                                      SAS Global O&G Domain

Acknowledgements

First and foremost, I like to acknowledge and thank the contribution of my colleagues at Intelligent Solutions, Inc. Razi Gaskari and Mohammad Maysami. Their hard work and commitment has played a major role in shaping the concepts that is presented in this book as "Shale Analytics".

I also would like to thank Professor Sam Ameri, Chair of Petroleum and Natural Gas Engineering Department at West Virginia University for his help and cooperation during the time that I was busy working on this book.

Some Chapters of this book were prepared in collaboration with the following co-authors:

Maher J. Alabboodi, West Virginia University
Dr. Soodabeh Esmaili, Devon Energy
Faegheh Javadi, Mountaineer Keystone
Dr. Amir Masoud Kalantari, University of Kansas
Dr. Mohammad Omidvar Eshkalak, University of Texas

Contents

Chapter 1
Introduction

Without Data, You Are Just another Person with an Opinion.
W.E. Deming (1900–1993).

When W.E. Deming uttered these words in 1980s, "world new nothing about the impact that shale will have in changing the energy landscape of the twenty-first century." Nevertheless, no other quotation can so vividly describe the state of our scientific and engineering knowledge of the physics of completion, stimulation, and interaction between induced and natural fractures during the oil and gas production from shale. Many well intentioned engineers and scientists that are involved in the day-to-day operations of shale wells develop the intuition required to make the best of what is available to them in order to increase the efficiency and the recovery from the shale wells. However, a full understanding of the physics and the mechanics of the storage and transport phenomena and the production operation in shale has remained elusive to a large extent.

There are many opinions and speculations on what exactly happens as we embark upon drilling, completing, and hydraulically fracturing shale wells. Much of the opinions and speculations are hardly ever supported by facts and data, and in the occasions that they are, it hardly ever transcends the anecdotal evidences that have been heard of, or been seen. However, the fact remains that as an industry, we collect a significant amount of data (field measurements—facts) during the operations that result in oil and gas production from shale. It is hard to imagine that these collected data do not contain the knowledge we need in order to optimize production and maximize recovery from this prolific hydrocarbon resource.

It took our industry professionals several years to come to the inevitable conclusion that our conventional modeling techniques that were developed for carbonate and coalbed methane formations cannot substitute our lack of understanding of the physics and the mechanics of completion and production from shale wells. But now that this fact has become self-evident, maybe the speculative resistance to the required paradigm shift to move to a data-driven solution can finally be overcome.

This book is dedicated to scratching the surface of all that is possible in the application of Petroleum Data Analytics to reservoir management and production

© Springer International Publishing AG 2017
S.D. Mohaghegh, *Shale Analytics*, DOI 10.1007/978-3-319-48753-3_1

operation of shale, what we have chosen to call "Shale Analytics." Here we demonstrate how the existing data collected from the development of shale assets can help in developing a better understanding of the nuances associated with operating shale wells. How to learn from our experiences in order to positively impact future operations. How to create a system of continuous learning from the data that is generated on a regular basis. In other words, using this technology we can make sure that every barrel of oil and every MCF of gas that is produced from our shale wells not only brings return to our investment, but also enriches our understanding of how this resource needs to be treated for maximum return.

1.1 The Shale Revolution

More than 1.8 trillion barrels of oil are trapped in shale in Federal lands in the western United States in the states of Colorado, Utah, and Wyoming, of which 800 billion is considered recoverable—three times the proven reserves of Saudi Arabia. The INTEK assessment for EIA found 23.9 billion barrels of technically recoverable shale oil resources in the onshore Lower 48 States. The Southern California Monterey/Santos play is the largest shale oil formation estimated to hold 15.4 billion barrels or 64 % of the total shale oil resources followed by Bakken and Eagle Ford with approximately 3.6 billion barrels and 3.4 billion barrels of oil, respectively [1].

Unlike the conventional oil and gas resources that are concentrated in certain parts of the world, shale resources (shale gas and shale oil) are widely distributed throughout the world. As such, it has the potential to democratize the energy production throughout the world. Figure 1.1 shows a map of the world along with the countries that have been studied in order to see if they have shale resources. Among the countries that have been studied (shown in white background color in Fig. 1.1), every one of them has been blessed with this organic rich resource play. Shale oil production has been estimated to extend to about 12 % of the global oil supply by 2035.[1] This may result in overall lower energy prices and consequently higher economic growth rate all over the world. Figure 1.2 shows the impact of shale on the United States' proven oil and gas reserves.

Until recently only 19 % of U.S. power generation was based on natural gas while more than 50 % was based on coal. Natural gas releases less than 50 % of the greenhouse gases than coal and just by switching from coal to gas considerable environmental impact can be expected with no impact on day-to-day life styles. Being able to reduce greenhouse gas emission by 50 % with no major impact on average people's life style or economic well-being would have been more like an environmentalist dream than reality only a decade ago.

---

[1]http://www.pwc.com/gx/en/industries/energy-utilities-mining/oil-gas-energy/publications/shale-oil-changes-energy-markets.html.

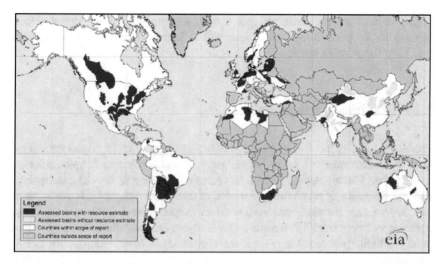

Fig. 1.1 Map of the shale resources in the world. Please note that countries identified by gray color were not assessed for their shale resources in this study performed for the Energy Information Administration (EIA). Most countries that were assessed (identified by *color white*) include shale reserves. *Source* (EIA—http://geology.com/energy/world-shale-gas/world-shale-gas-map.gif)

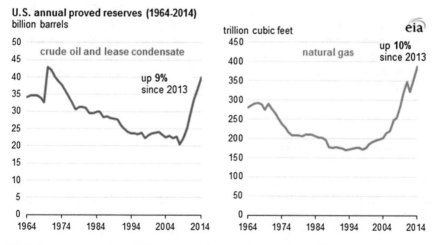

Fig. 1.2 Impact of shale on U.S. proven oil and gas reserves. *Source* EIA

Natural gas was not being considered as a serious alternative to coal since coal was abundant and cheap. The supply crunch of natural gas, which resulted in construction of facilities to import Liquefied Natural Gas (LNG) from abroad, is now history. The LNG facilities are now being modified to export rather than import natural gas. Once natural gas can be exported and is no longer a local commodity, its price in the United State will be impacted; however, its abundance will keep the price in check.

1.2 Traditional Modeling

Petroleum industry has changed significantly since production of oil and natural gas from shale has become a profitable endeavor. Shale as a source for production of hydrocarbon is a revolutionary idea that has now become reality. The technology that has realized this revolution is innovative and game changing. However, as an industry we are still using old, conventional technologies to analyze, model, and optimize recovery from this resource. Our analyses and modeling efforts up to now have had limited success. Many engineers and managers in the industry now openly admit that numerical simulation and RTA (Rate Transient Analysis) add very little value to their operation.

This could have been foreseen (and actually was discussed and published in 2013 [2, 3]) based on the essence of these technologies and how they model the network of natural fractures, induced hydraulic fractures, and the way they are coupled and interact in shale. When it comes to production from shale using long horizontal wells that are hydraulically fractured in multiple stages, these conventional technologies are too simplistic and are not capable of realistically modeling the physics (as much of it as we understand) of the problem. Therefore, they make unreasonable simplifying assumptions to a degree that make their use all but irrelevant. However, in the absence of any other widely accepted technology as an alternative for modeling the storage and transport phenomena in shale, these technologies flourished in the past several years.

1.3 A Paradigm Shift

"Paradigm Shift," a term first coined by Kuhn [4], constitutes a change in basic assumptions within the ruling theory of science. According to Kuhn, "A paradigm is what members of a scientific community, and they alone, share" [5]. Jim Gray, the American computer scientist, who received the Turing Award for his seminal contribution to computer science and technology coined the term "Fourth Paradigm" of science referring to data-driven science and technology [6].

In his classification the first paradigm of science (thousands of years ago) was about empirically describing the natural phenomena. The second paradigm of science (last few hundred years) was the theoretical branch of science where we use models and generalizations to describe nature, examples of which are Kepler's Law, Newton's Law of Motion, and Maxwell's Equations. When the theoretical models grew too complicated to solve analytically, the third paradigm of science was born (last few decades). This was the computational branch of science that includes simulating complex phenomena. Today, the fourth paradigm is the data-Intensive, data exploration, or "eScience."

This paradigm unifies theory, experiment, and simulation where data is captured by instruments or generated by simulator and is processed by software and information/knowledge stored in computer. Scientists analyze database/files using data management tools and statistics. Based on Jim Gray's view "The World of Science Has Changed. There is No Question About it. The techniques and technologies for data-intensive science are so different that is worth distinguishing data-intensive science from computational science as a new, fourth paradigm for scientific exploration."

Petroleum Data Analytics, which is the application of data-driven analytics in the upstream oil and gas, represents the paradigm shift articulated by Gray [6]. Petroleum Data Analytics helps petroleum engineers and geoscientist build predictive models by learning from hard data (field measurements). It has proven to be the alternative to traditional technologies. Data-driven analytics is gaining popularity among engineers and geoscientists as it proves its predictive capabilities and as more solutions in the form of software products[2] surface. The main objective of this book is to cover the state of the art in application of data-driven analytics in analysis, predictive modeling, and optimization of hydrocarbon production from shale formations. We have named this technology "Shale Analytics."

---

[2]IMprove is a software application by Intelligent Solutions, Inc. that helps user to analyze hard data (field measurements), build and validate predictive models, and perform post-modeling analyses for production optimization from shale formations.

Chapter 2
Modeling Production from Shale

Mitchell[1] and his team of geologists and engineers began working on the shale challenge in 1981, trying different combinations of processes and technologies before ultimately succeeding in 1997 with the use of a "slick-water" frac that made Barnett Shale economical to develop and in turn changed the future of the US natural gas industry [7]. Continuing on Mitchell's success progress followed a path that included horizontal wells, multi-cluster, multistage, hydraulic fracturing of horizontal wells and pad drilling, and the rest is history.

The success in overcoming the technical difficulties to unlock the huge potentials of oil and gas production from shale is very much tied to an integration of long lateral horizontal drilling, coupled with multistage, multi-cluster hydraulic fracturing that initiates new fractures while activating a system of natural fracture networks in shale. This system of highly permeable conduits introduces the highly pressurized shale formation to a lower pressure, causing a pressure gradient that triggers the flow of oil and natural gas to the surface at very high rates, albeit, with steep decline. Looking at the list of most important technological innovations that have made production from shale possible one can see the challenges that are associated with understanding and modeling and eventually optimizing the production of hydrocarbon from shale. So let us examine what are the realities that we are aware of, and how we go about modeling them in our traditional techniques.

First and foremost is the fact that shale is naturally fractured. The natural fractures are mostly sealed by material that have precipitated in them and/or reside in the natural fracture as a result of chemical reactions of the material present in the fractures throughout the geologic time. The only way we can detect presence of the system of natural fractures in the area being developed is by detecting (using logs, video, etc.) the presence of the intersection of these natural fractures with the wellbore. Even when we are successful in detecting the presence of natural fractures, we only have indications about their width (openings) and direction. We

---

[1]Mitchell Energy and Development. He sold his company to Devon Energy in 2002 in a deal worth $3.5 Billion.

© Springer International Publishing AG 2017
S.D. Mohaghegh, *Shale Analytics*, DOI 10.1007/978-3-319-48753-3_2

cannot tell how far beyond the wellbore they extend. The rest is actually (educated) guess work. To model this system of natural fracture networks to be used in our numerical simulation models (i.e., when we actually bother to be this detailed in our analysis), we "estimate" parameters such as direction and density of the major and minor fracture networks and generate them stochastically. In other words, there is not much that we can actually measure and what we model has little to do with realities that can be measured. Once the presence of the system of the natural fracture networks is modeled, then it can be made to be used during the flow modeling.

The reality is that while the system of natural fracture networks is under extensive pressure and stress, it still presents a network of sealed conduits that is vulnerable to failure (opening) prior to the matrix of the rock, once an external force finds its way into the rock to counter the overburden pressure and the in situ stresses. This breaking of the rock (fracking) is caused by hydraulic pressure exerted during the hydraulic fracturing. During hydraulic fracturing large amount of "slick water" (or other fracking fluids) is injected into the rock at pressures that is above the natural in situ pressure in order to break the rock. The crack that is initiated in the rock is extended by continuing the injection of the fluid. The rock will break at its weakest points that are usually the system of natural fracture networks along with other places in the fabric of the rock where there are structural vulnerabilities.

All previous (theoretical and modeling) work on hydraulic fracturing is concentrated on propagation of hydraulic fractures in nonnaturally fractured formations, specifically when the hydraulic fracturing is to be modeled (coupled) in a numerical reservoir simulation that will be used to model fluid flow to the wellbore and eventually to the surface. As such, all such models assume a so-called a "penny-shaped" (or modified penny shaped) hydraulic fracture propagation into the formation. Figure 2.1 shows examples of traditional modeling of hydraulic fractures. Well, although we may not know what is the shape of the hydraulic fracture propagation in a shale formation, it would be very hard to find a reservoir engineer or completion engineer that would be naïve enough to think that hydraulic fractures in shale is anything remotely like a penny (Fig. 2.1). In other words, although the shape and the propagation of the hydraulic fractures in shale are unknown, it definitely is not "penny-shaped".

These gross simplifying assumptions that are used during the modeling of hydraulic fracturing in shale demonstrate that we have been modeling shale using a "Pre-Shale" technology. Coining the phrase "Pre-Shale" technology aims to emphasize the combination of technologies that are used today in order to address the reservoir and production modeling of shale assets. In essence, almost all of the technologies that are used today for modeling and analyses of hydrocarbon production from shale were developed to address issues that had originally nothing to do with shale. As the "shale boom" started to emerge these technologies are revisited and modified in order to find their application in shale. For example, the way we numerically model fluid flow in, and production from, shale is essentially a combination of what our industry has devised to better understand, address, and

Fig. 2.1 Traditional penny-shaped model of hydraulic fracturing **a** nonmodified, **b** shape of the fracture is modified due to nonuniform stress

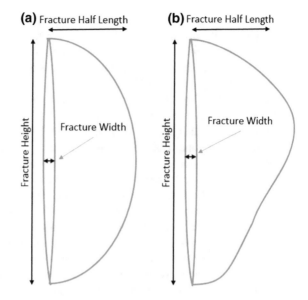

model carbonate (Discrete Fracture Networks) and coalbed methane (diffusion of gas through the matrix via concentration gradient). This has given rise to today's numerical simulation formulation for shale that can be summarized as "Carbonate + CBM = Shale." Technologies such as wellbore image logs and microseismic are not much different and can fit this definition as well.

Most of the analytical solutions to the flow in the porous media as well as other simplified solutions may also be included in the "Pre-Shale" technology category. Technologies such as Decline Curve Analysis, Rate Transient Analysis, Volumetric calculation of reserves and material balance calculations may be categorized as the "Pre-Shale" technologies. Of course some of these techniques are fundamental enough to find application to shale (like material balance) but their full applicability is still a function of better understanding of the storage and flow mechanisms in shale and that is yet to be solidified.

2.1   Reservoir Modeling of Shale

Since there seems to be plenty of "Unknown Unknowns" and a certain number of "Known Unknowns" when it comes to storage and fluid flow in shale, it may not be a bad idea to start with some "Known Facts" and see if we can come up with some general ideas that enjoy wide acceptance among the professionals in the industry. *Fact Number One* is that "Shale is Naturally Fractured." This is a fact that hardly anyone will dispute. A quick survey of the papers published on the reservoir simulation and molding of shale (or any other analysis regarding hydrocarbon production from shale) shows that almost everyone starts with the premise that shale is naturally fractured. Please note that at this point in the book the nature, characteristics, and distribution of the natural fractures in shale are not being considered. Just the fact that shale contains a vast network of natural fractures is the essence of the Fact Number One.

Fact Number Two that seems to have been widely accepted is that "Hydraulic (induced) Fractures Will Open (activate) Existing Natural Fractures." Many recent modeling techniques (few of them being cited here) start with such a premise in order to map the complexities of induced fracture in shale. Even if we believe that hydraulic fracture will create new fractures in shale (which seems to be a fact as well), it would be very hard to argue against the notion that it can and will open existing natural fractures in shale. This is due to the fact that existing natural fractures provide a path of least resistance to the pressure that is imposed on the shale during the process of hydraulically fracturing the rock.

Unfortunately, it seems that here is where the "Known Facts" that are widely accepted among most scientists and engineers, comes to an end. Almost every other notion, idea, or belief, is faced with some sort of a dispute by some along with reasonably strong arguments "for" and "against" them.

2.2 System of Natural Fracture Networks

Reservoir development is impacted by natural fractures in three ways. First, natural fractures are planes of weakness that may control hydraulic fracture propagation. Second, high pressures from the hydraulic fracture treatment may cause slip on natural fractures that increases their conductivity. Third, natural fractures that were conductive prior to stimulation may affect the shape and extent of a well's drainage volume [8].

Natural fractures are Digenetic fractures and/or tectonic fractures. Natural fractures are mechanical breaks in rocks, which form in nature, in response to lithostatic, tectonic and thermal stress, and high fluid pressure. They occur in a variety of scales and with high degree of heterogeneity [9].

The most common technique for modeling the System of Natural Fracture Networks (SNFN) is to generate them stochastically. The common practice in carbonate and some clastic rocks is to use Borehole Image Logs in order to characterize the SNFN at the wellbore level knowing that such characterization is only valid a few inches away from the wellbore. These estimates of SNFN characteristics are then used for the stochastic generation of the discrete fracture networks throughout the reservoir. Parameters such as mean and standard deviation of fracture orientation, form of fracture length distribution, averages for fracture length, aperture (width), density of center points and relative frequency of terminations are among the characteristics that are needed (guessed or estimated) so that the stochastic algorithms can generate a given System of Natural Fracture Networks.

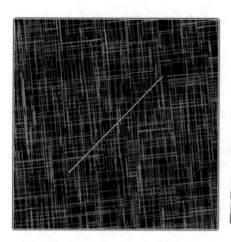 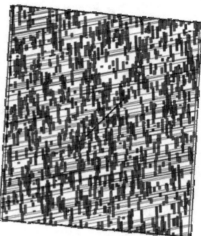

Fig. 2.2 System of Natural Fractures Networks (SNFN) generated using stochastic techniques

Sometimes such exercise is performed in multiple sets, changing the afore-mentioned parameters in order to generate multiple sets of networks to resemble some of the observed characteristics in the outcrops. Figure 2.2 displays typical networks of natural fractures that are generated using stochastic techniques. For the purposes of this chapter, we name this type of generation of natural fracture networks, the "Conventional SNFN" to distinguish its characteristics, and the conse-quences of its use and implementation, from the potential SNFN that we postulate happening in shale as "Shale SNFN."

System of Natural Fracture Networks models have many advantages over conventional Dual Porosity (DP) approaches, especially in heterogeneous reservoirs where the dominant flow mechanism is through the network of fractures rather than the reservoir matrix. The SNFN approach is based on the stochastic modeling concept and therefore, every realization of the System of Natural Fracture Networks will produce different results. As such, SNFN-type modeling is not a direct com-petitor to DP reservoir modeling. Rather, it provides an additional insight into the potential variability of production histories [10].

Idea of SNFN is not new. It has been around for decades. Carbonate rocks and some clastic rocks are known to have networks of natural fractures. Developing algorithms and techniques to stochastically generate SNFN and then couple them with reservoir simulation models was common practice before the so called "shale revolution." Most recently a number of investigators have attempted to model production from shale by making effective use of the SNFN and its interaction with the induced fractures.

Li et al. [11] proposed a numerical model that integrates turbulent flow, rock stress response, interactions of hydraulic fracture propagation with natural fractures, and influence of natural fractures on formation's Young's modulus. They postulated

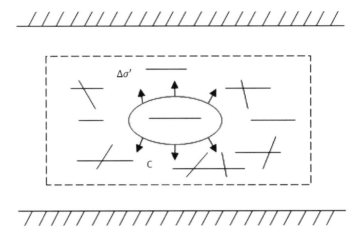

Fig. 2.3 Preexisting Natural Fracture distribution in shale [11]

that the preexisting natural fractures in shale formation complicate hydraulic fracture propagation process and alter its Young's modulus. Their preliminary numerical results illustrate the significant differences in modeling hydraulic fracture propagation in comparison with current models that assume laminar flow in hydraulic fracture process. They conclude that length and density of natural fracture have significant impact on formation Young's modulus, and interactions between hydraulic fracture and natural fractures create complex fracture network.

Figures 2.3 and 2.4 [11] clearly show the SNFN used in the literature that we have named Conventional SNFN. The impact of the nature and distribution of the SNFN in the overall performance of the well and specifically in the propagation of the hydraulic fracture in shale has been emphasized in the literature.

Other authors [12] have presented simulation results from complex fracture models that show stress anisotropy, natural fractures, and interfacial friction play critical roles in creating fracture network complexity. They emphasize that decreasing stress anisotropy or interfacial friction can change the induced fracture geometry from a bi-wing fracture to a complex fracture network for the same natural fractures. The results presented illustrate the importance of rock fabrics and stresses on fracture complexity in unconventional reservoirs.

Figure 2.2 [12] shows that the natural fracture network that they have considered in their development is very much the same as mentioned in other papers when it comes to propagation of hydraulic fractures in shale and its interaction with the natural fractures, a system of natural fractures that we have chosen to call the Conventional SNFN.

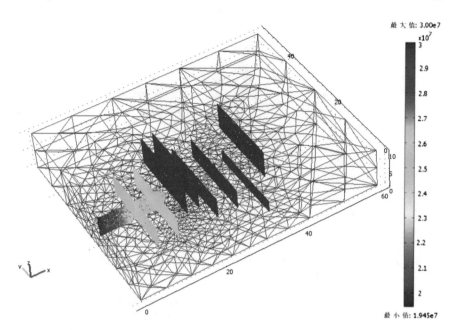

最 大 值: 3.00e7

x10⁷

Fig. 2.4 Hydraulic Fracture propagation distribution at time-step 20 [11]

Recent petroleum engineering literature is full of similar examples. They have two common themes

1. The preexisting system of natural fracture networks in shale formations plays a dominant role in determining the propagation path of the induced hydraulic fracture and consequently determines the degree of productivity of hydrocarbon producing shale wells,
2. Conventional SNFN is the only form of network of natural fractures that is considered in shale formations.

While the first point is well established and commonly accepted among most of the engineers and scientists, and is accepted by the author, the second point should not be taken so lightly. Author would like to propose an alternative to this commonly held belief that the network of natural fractures in shale can be categorized as what we have called in this manuscript to be the Conventional SNFN.

2.3 System of Natural Fracture Networks in Shale

Above examples demonstrate that although different scientists and researchers attempted to find better and more efficient ways to address the propagation of hydraulic fractures in shale, all of them have one thing in common. They all use the

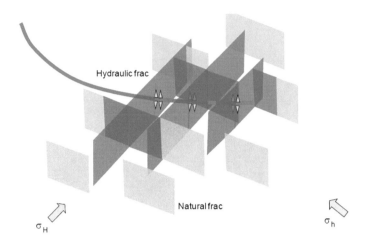

Fig. 2.5 Diagram of hydraulic fracture network and preexisting natural fractures [12]

legacy definition and description of SNFN. As was shown above, this legacy description includes a network of natural fractures that exist in the fabric (matrix) of the porous medium and it is mainly characterized by random occurrences, length, aperture, and intersections and is described by J1 and J2 type fractures (Figs. 2.2, 2.3, 2.4 and 2.5).

But what if this legacy definition and description of network of natural fracture that is essentially borrowed from carbonate rocks and is an indication of our lack of understanding and ability to visualize and measure them in the matrix, is not applicable to shale? What if the network of natural fractures in shale has a completely and fundamentally different nature, structure, characteristics, and distribution than what is commonly used in all of our (commercial, academic, and in-house) models?

2.4   A New Hypothesis on Natural Fractures in Shale

What is the general shape and structure of natural fractures in Shale? Is it closer to a stochastically generated set of natural fracture with random shapes that has been used for carbonates (and sometimes clastic) formations? Or is it more like a well-structured and well-behaved network of natural fracture that have a laminar, plate-like form, examples of which can be seen in the outcrops such as those shown in the Fig. 2.6

Shale is defined as a fine-grained sedimentary rock that forms from the compaction of silt and clay-size mineral particles that we commonly call "mud." This

Fig. 2.6 Examples of natural fractures in shale that is clearly observable from the outcrops

composition places shale in a category of sedimentary rocks known as "mud-stones." Shale is distinguished from other mudstones because it is fissile and laminated. "Laminated" means that the rock is made up of many thin layers. "Fissile" means that the rock readily splits into thin pieces along the laminations.[2]

If such definitions of the nature of shale is accepted and if the character of network of natural fractures in shale is as it is observed in the outcrops and depicted in the diagram of Fig. 2.7, then many questions must be asked, some of which are

(a) How would such characteristics of the network of natural fractures impact the propagation of the induced hydraulic fractures in shale?
(b) How would the production characteristics of shale wells are impacted by this potentially new and completely different way of propagation of the induced hydraulic fractures (as compared to how we model them today).
(c) What are the consequences of these characteristics of natural fractures on the short and long-term production from shale?
(d) How would this impact our current models? And finally,
(e) What can it tell us about the new models that need to be developed?

Obviously, there are many more questions that can be asked. Here we postulate that such definition of the system of natural fractures in shale is similar to those shown in Figs. 2.6 and 2.7. We then try to hypothesize the consequences of such assumptions.

---

[2]http://geology.com/rocks/shale.shtml.

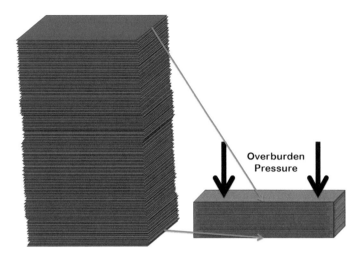

Fig. 2.7 Schematic of the nature of the SNFN in shale outcrops and its potential shape when subjected to overburden pressure

2.5 Consequences of Shale SNFN

To address some of the questions posed above, one needs to observe that if the natural fracture network in shale is indeed anything like what is suggested in this chapter then we may have to go back to the drawing board and start the development of our shale models from scratch. Given the thin nature of the plates one must consider the density of the plates, or density of natural fractures per inch of formation thickness. While fluid flow in matrix or fabric of the shale remains the territory of diffusion of gas through solids, modeling of the flow through propped open natural fractures and interaction between natural fractures and the rock matrix may no longer be efficiently modeled as flow through porous media. May be flow through parallel plates coupled with diffusion is a more robust manner of modeling.

On the other hand, this new way of thinking about Shale SNFN may enable us to provide a reasonable answer to the large amount of hydrocarbon that is produced upon hydraulically fracturing the shale and can substitute the unrealistic and in some cases even humorous notion that the hydraulic fractures in shale are penny shaped (may be somewhat deformed) and can be modeled in the same manner that we used to model the hydraulic fracture propagation in carbonate and clastic formations.

Previously we mentioned that it is widely accepted that (a) shale is naturally fractured, and (b) the induced hydraulic fracture tend to first open the existing natural fracture. If the two above mentioned facts are accepted, then the natural next step may be to discuss the shape, the characteristics, and the distribution of the natural fracture networks in shale.

Almost all the published papers assume that the natural fracture networks are stochastic in nature and therefore must be modeled as such. Furthermore, these assumptions inherently include only vertical fractures in the form of J1 and J2, etc. They follow by identifying a series of statistical characteristics that will be used in a variety of algorithms that will generate natural fracture networks. Once generated, the natural fracture networks are treated in many different ways in order to contribute to the reservoir modeling of shale assets. Effect and impact of these natural fracture networks are approximated analytically in some studies, while they are solved using an elaborate system of equations in other studies. Some have opted to use the natural fracture networks in order to identify the complex growth of hydraulic fractures. The authors observe that in all these cases the shape, the characteristics, and the distribution of the natural fracture networks in shale are common and include only vertical fractures in the form of J1 and J2, etc. The inevitable question is: "are we using these types of shape, characteristics, and distribution because we are able to readily model them in our reservoir simulation codes, or that we have coded such shapes, characteristics, and the distributions, because this is what we believe is happening"?

We start by posing a set of questions

1. What is the most probable shape for the network of natural fractures in shale?
2. When we hydraulically fracture shale, is it possible that we are opening the existing horizontal and plate-like natural fractures, before, or during creation of other fractures ?[3]
3. What are the consequences of opening these well-behaved, horizontal plate-like natural fractures in shale during the hydraulic fracturing?
4. Are the existing simulators adequate for modeling the production from shale wells, if indeed the above hypotheses are correct?

The answer may be revealed if the question is asked in a different fashion. If the dominant natural fracture networks in shale are horizontal (instead of vertical as shown in Fig. 2.2, 2.3, 2.4 and 2.5) can our current reservoir simulation models handle them? Imagine a vast, massive network of horizontal natural fractures with solid plates no thicker than 1–2 mm (essentially a stack of cards) that can be opened upon hydraulic fracturing and can contribute to flow.

This type of model provides a very large porosity that initially (prior to hydraulic fracturing) is not necessarily connected (or is only connected locally and in limited scope). This vast network of natural fractures is opened and become connected upon hydraulic fracturing which then creates substantial permeability. Furthermore, the very thin nature of the solid (very tight) rock plates that are themselves easier to crack upon losing their original calcite support (though giving rise to potential J1, J2 type fractures) are the medium for possible diffusion of trapped hydrocarbon

---

[3]This notion, of course, has the obvious issue with the magnitude of stresses at different direction. It seems that the overburden pressure (vertical stress) is not the minimum stress. Therefore, maybe, the horizontal/laminated fractures are not the first set of fractures that open. However, totally dismissing theses natural fractures seem to be an oversight that needs to be addressed.

(in addition of the hydrocarbon in the dominant horizontal natural fracture networks that are released upon opening) to support continued production.

2.6   "Hard Data" Versus "Soft Data"

As we move forward with explaining use of traditional technologies in modeling reservoir and production from shale it is important to explain and distinguish between "Hard Data" and "Soft Data." This is necessary since in this book we will demonstrate that Shale Analytics is a technology that uses "Hard Data" to model production from shale while most of the traditional technologies use "Soft Data."

"Hard Data" refers to field measurements. This is the data that can readily be, and usually is, measured during the operation. For example, in hydraulic fracturing variables such as fluid type and amount, proppant type and amount, injection, breakdown and closure pressure, and injection rates are considered to be "Hard Data." In most shale assets "Hard Data" associated with hydraulic fracturing is measured and recorded in reasonable detail and are usually available. Table 2.1 shows a partial list of "Hard Data" that is collected during hydraulic fracturing as well as a list of "Soft Data" that is used by reservoir engineers and modelers.

In the context of hydraulic fracturing of shale wells, "Soft Data" refer to variables that are interpreted, estimated, or guessed. Parameters such as hydraulic fracture half length, height, width and conductivity cannot be directly measured. Even when software applications for modeling of hydraulic fractures are used to estimate these parameters, the gross limiting and simplifying assumptions that are made, such as well-behaved penny like double wing fractures (see Fig. 2.1), renders the utilization of "Soft Data" in design and optimization of frac jobs irrelevant.

Table 2.1   Examples of hard versus soft data for hydraulic fracture characteristics

Hard Data	Soft Data
Fluid types	Hydraulic Fracture Half Length
Fluid amounts (bbls)	Hydraulic Fracture Width
Pad volume (bbls)	Hydraulic Fracture Height
Slurry volume (bbls)	Hydraulic Fracture conductivity
Proppant types	Stimulated Reservoir Volume:
Proppant amounts (lbs)	• SRV height
Mesh size	• SRV Width
	• SRV length
Proppant Conc. (Ramp Slope)	• SRV Permeability
Max. Proppant Concentration	
Injection Rate	
Injection Pressure: • Average Inj. Pressure • Breakdown Pressure • ISIP • Closure Pressure	

Another variable that is commonly used in the modeling of hydraulic fractures in shale is Stimulated Reservoir Volume (SRV). SRV is also "Soft Data" since its value cannot be directly measured. SRV is mainly used as a set of tweaking parameters (dimensions of the Stimulated Reservoir Volume as well as the permeability value or values that are assigned to different parts of the stimulated volume) to assist reservoir modelers in the history matching process.

2.7 Current State of Reservoir Simulation and Modeling of Shale

Since reservoir simulation and modeling of shale formations became a task to be tackled by reservoir engineers, the only available option, and therefore the solution that has been presented, has been a modified version of existing simulation models. These modifications are made so that the existing simulators can mimic the storage and flow characteristics in shale. Although our information regarding the required characteristics of a shale-specific simulation models were quite limited, the possible choices in using the existing simulators meant the inclusion of a combination of algorithms such as discrete fracture networks, dual porosity and stress dependent permeability, as well as adding concentration driven Fickian flow and coupling it with Langmuir's isotherms. However, our inability to model hydraulic fractures and its nonuniform propagation in a naturally fractured system did not stop us from going forward with the business of modeling.

In other words, our choices, especially at the start of this process, were quite limited. Probably the main reason was that the industry was, and still is, in need of tools that can help in making the best possible decision during the asset development process. Although some interesting work has been performed, especially in the area of transport at the micro-pore level, they have not yet found their way into the popular simulation models that are currently being used by the industry.

The current state of reservoir modeling technology for shale uses the lessons learned from modeling naturally fractured carbonate reservoirs and those from coalbed methane (CBM) reservoirs in order to achieve its objectives. The combination of flow through double porosity, naturally fractured carbonate formation, and concentration gradient driven diffusion that is governed by Fick's law integrated with Langmuir isotherms that controls the desorption of methane into the natural fractures, has become the cornerstone of reservoir modeling in shale.

Most of the competent and experienced reservoir engineers and modelers that the author has communicated with regarding this issue recognize the shortcomings of this approach when applied to shale. Nevertheless, all agree that this is the best option that is currently available when we attempt to numerically model fluid flow through shale. While most of the recent reservoir simulations and modeling of shale have the above approach in common, they usually vary on how they handle the massive multi-cluster, multistage hydraulic fractures that are the main reason for economic oil and gas production from shale reservoirs.

The presence of massive multi-cluster, multistage hydraulic fractures only makes the reservoir modeling of shale formation more complicated and the use of current numerical models even less beneficial. Since hydraulic fractures are the main reason for economic production from shale, modeling their behavior and their interaction with the rock fabric, becomes one of the most important aspects of modeling storage and flow in shale formations. Therefore, the relevant question that should be asked is: How do the current numerical reservoir simulation models handle these massive multi-cluster, multistage hydraulic fractures?

When the dust settles and all the different flavors of handling multi-cluster, multistage hydraulic fractures in reservoir modeling are reviewed, all the existing approaches can be ultimately divided into two distinct groups. The first is the Explicit Hydraulic Fracture (EHF) modeling method, and the second is known as Stimulated Reservoir Volume (SRV).[4] We will briefly discuss these techniques later in this chapter.

Before examining some details of the EHF and SRV techniques, it must be mentioned that there are a couple of other techniques that have been used in order to model and forecast production from shale wells. These are Decline Curve Analysis (DCA) and Rate Transient Analysis (RTA). These two methods are quite popular among practicing engineers for their ease of understanding and use.

2.7.1 Decline Curve Analysis

Decline Curve Analysis (DCA) is a well-known and popular technology in our industry. The popularity of DCA is due to its ease of use (and in many cases it can be, and is, easily misused). When applied to shale wells DCA has many shortcomings. Several authors [6, 13–17] have come up with interesting techniques to overcome some of the well-known shortcomings of DCA, but nevertheless, many facts remain that make the use of Decline Curve Analysis suboptimal.

One of the major criticisms of Decline Curve Analysis is its lack of sensitivity to major physical phenomena in shale wells that has to do with the fluid flow, the hydraulic fracture, and the reservoir characteristics. In cases like Marcellus and Utica shale reservoirs where short periods of production are available, the use of Decline curve Analysis becomes increasingly problematic.

When it comes to hydrocarbon production from shale, there is a major flaw in the application of decline curve analysis, whether it be the ARP's original formulation or other flavors that have emerged in the recent years [18–20]. Limitations associated with decline curve analysis are well known and have been discussed

---

[4]Some have chosen to use alternative nomenclature such as Estimated Stimulate Volume (ESV) or the Crushed Zone, but the idea behind them is all the same.

comprehensively in the literature. Nevertheless, most of the limitations that have to do with specific flow regimes or nuances associated with operational inconsistencies have been tolerated and clever methods have been devised to get around them. However, when it comes to analysis of production data from shale, a new set of characteristics, that may not have been as dominant in the past, stands out that seriously undermine the applicability of any and all production analysis techniques that rely on traditionally statistics-based curve fitting techniques (including DCA).

What is new and different about production from shale is the impact and the importance of completion practices. There should be no doubts in anyone's mind that a combination of long laterals with massive hydraulic fractures is the main driver that has made economic production from shale a reality. So much so that professionals in the field have started questioning the impact and the influence of reservoir characteristics and rock quality in the productivity of shale wells [21]. While the importance of reservoir characteristics is shared between conventional and unconventional wells, design parameters associated with the completion practices in wells producing from shale are the new and important set of variables that create the distinction with wells in conventional resources. In other words, completion design parameters introduce a new set of complexity to production behavior that cannot be readily dismissed, and they are being completely dismissed and overlooked anytime decline curve analysis are used in shale.

Therefore, there is a new set of inherent and implicit assumptions that are associated with production data analyses in shale when methods such as decline curve analysis are used. In previous cases (non-shale) production is only a function of reservoir characteristics while the human involvements in form of operational constraints and completion design parameters played minimal roles. In shale, completion design practices play a vital role. By performing traditional statistics-based production data analysis, we are assuming that reasonable or optimum or may be even consistent completion practices are common in all the wells in a given asset. Production and completion data that have been thoroughly examined in multiple shale plays, clearly point to the fact that such assumptions are indeed invalid and may prove to be quite costly for the operators.

2.7.2 Rate Transient Analysis

Rate Transient Analysis (RTA) is a clever technology [22–27] that approximates the essence of reservoir simulation and modeling using a series of analytical and graphical (plotting routines) approaches. RTA's ease of use and consistency of results are among its strong points. On the other hand, RTA suffers from the same problems as numerical reservoir simulation and modeling does, since almost all of its approaches, especially when it forecasts production, mimics those of numerical modeling.

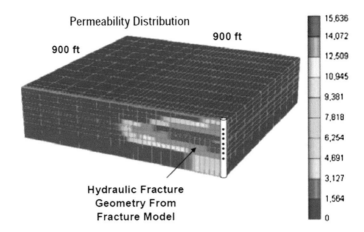

Fig. 2.8 Example of explicit hydraulic fracture (EHF) modeling [28]

2.8   Explicit Hydraulic Fracture Modeling

When compared with other traditional techniques, Explicit Hydraulic Fracture (EHF) modeling is the most comprehensive, complex, and tedious (as well as the most robust) approach for modeling the impact of hydraulic fracturing during numerical simulation of production from shale (example shown in Fig. 2.8). The Explicit Hydraulic Fracture (EHF) modeling technique of reservoir simulation and modeling of shale wells couples three different technologies (software applications [hydraulic fracture modeling software, geological modeling software, and numerical reservoir simulation software]) and includes the following steps:

1. *Modeling the impact of the hydraulic fracture*; during this step each cluster of hydraulic fracture is modeled individually using independent hydraulic fracture simulation software applications such as MFrac,[5] FracPro,[6] etc. These models use the frac job characteristics (recipe) such as fluid and proppant amount and rate of injection, along with some reservoir characteristics and stresses, and calculate the characteristics of an idealized hydraulic fracture. It is important to note that most of the time the reservoir characteristics (including the stresses) needed as input to these models are not available and are guessed (assumed) so that these models can be used.

   Since these models assume a well-behaved penny-shaped hydraulic fracture (albeit a deformed penny from time to time—see Fig. 2.1), the characteristics they calculate are fracture half length, fracture height, fracture width, and fracture conductivity. This process is repeated for every single cluster of hydraulic fractures. This means that in some cases up to 60 or 70 hydraulic

---

[5]Meyer Fracturing Software, a Baker-Hughes Company, www.mfrac.com.

[6]Carbo Ceramics, http://www.carboceramics.com/fracpropt-software/.

fracture clusters per well (about three clusters per stage) need to be modeled independently.

2. **Developing a geological model**; as in all other serious reservoir simulation and modeling exercises, developing a geological model is a necessary step in the numerical modeling of production from shale. During this step all the geological, petrophysical and geophysical information available to the modeling team is used to develop a reasonably detailed geological model. Even for a single well model this process may generate a detail multimillion grid block geological model. Usually data from all the available wells are used to generate the structural map and volume that is then discretized and populated with appropriate data based on availability. This process is usually performed using a geological modeling software application, several of which are currently available in the market and are extensively used during the modeling process. Inclusion of Discrete Natural Fracture Network (DNF) in the modeling process is usually performed during this step. The common approach is to develop the DFN using statistical means and then use analytical or numerical technics to incorporate the impact of the develop DFN into the existing grid block system developed during the construction of the geocellular model.

3. **Incorporation of frac characteristics in the geological model**; in order to incorporate the hydraulic fracture characteristics into the geological model, first the wellbore must be included. Upon inclusion of the wellbore, all the calculated characteristics from step 1 (hydraulic fracture impact such as fracture half length, fracture height, fracture width, and fracture conductivity), are imported into the geological model (mentioned in step 2). This is a rather painstaking process through which the grid system developed during geological modeling is modified in order to be able to accommodate the hydraulic fracture characteristics. Usually a local grid refinement process is required (both horizontally as well as vertically) for this process. The result is usually a detail model that includes a large number of grid blocks. When building a model that includes multiple pads and wellbores this process may take a long time.

   Due to the detailed nature of the model, the computational cost of such models is very high. This fact makes full field modeling of shale assets, impractical (in this book we introduce a solution for this specific problem—Chapter Nine—using data-driven analytics). That is the main reason behind the fact that the overwhelming number of numerical simulation studies conducted on shale formations is single well models. From time to time one may find studies that are performed on a pad of multiple horizontal wellbores rather than a single well, but such studies are few and far between.

4. **Completing the base model**; using numerical reservoir simulation software application. Completion of the base model usually requires some up-scaling and incorporation of operational constraints. Identification and incorporation of appropriate outer boundary conditions and making a first run to check for convergence are among the other steps that need to be taken for the completion of the base model.

5. *History matching the base model*; once the base model is completed and runs properly, the difference between its results and the observed measurements (e.g., production rates from the field) indicates the proximity of the model to where it needs to be. During the history matching process, geological and hydraulic fracture characteristics are modified until an acceptable history match is achieved.

6. *Forecasting production*; the history matched model is executed in the forecast mode in order to predict future production behavior of the shale well.

A survey of most recent publications shows that many modelers have selected not to use the Explicit Hydraulic Fracture (EHF) modeling methodology. This may be attributed to degree of detail that goes into building and then history matching an Explicit Hydraulic Fracture (EHF) model for shale wells. The amount of time it takes to complete the above steps for a moderate number of wells can be quite extensive. Imagine trying to build a full field model where tens or hundreds of wells are involved. The size of such a model can (and usually does) make the development and history matching process computationally impractical.

2.9   Stimulated Reservoir Volume

The second technique for modeling production from shale wells is known as Stimulated Reservoir Volume (SRV) modeling technique. Stimulated Reservoir Volume (SRV) modeling technique is a different and much simpler way of handling the impact of massive multi-cluster, multistage hydraulic fractures in numerical reservoir simulation and modeling. Using SRV instead of EHF can expedite the modeling process by orders of magnitude. This is due to the fact that instead of meticulously modeling every individual hydraulic fracture, in this method the modeler assumes a three dimensional volume around the wellbore with enhanced permeability as the result of the hydraulic fractures (see Figs. 2.9 and 2.10). By modifying the permeability and dimensions of the Stimulated Reservoir Volume (SRV), the modeler can now match the production behavior of a given well in record time.

The first question that comes to mind upon understanding the impact of the Stimulated Reservoir Volume on production is how one would calculate, or more accurately, estimate, the size of the Stimulated Reservoir Volume. Given the fact that Stimulated Reservoir Volume results from hydraulic fractures, the next question that comes to mind is whether the SRV is a continuous medium or it has discrete characteristics for each hydraulic fracture and whether or not these discrete volumes are connected to one another. Furthermore, how are the aspect ratios (ratio of height, to width and to length) of the Stimulated Reservoir Volume determined?

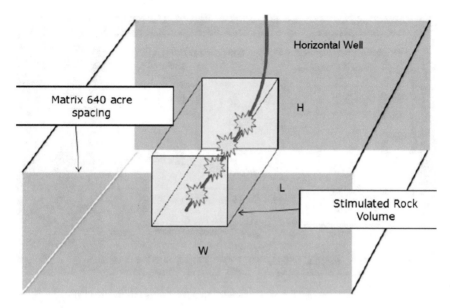

Fig. 2.9 Example of stimulated reservoir volume [29]

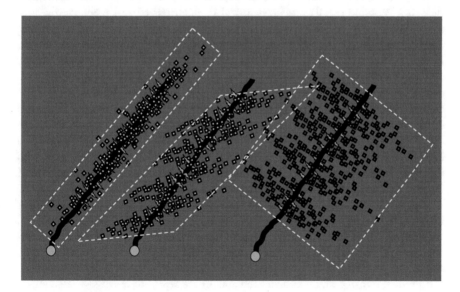

Fig. 2.10 Example of stimulated reservoir volume [30]

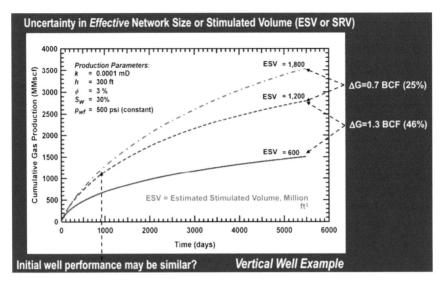

Fig. 2.11 Sensitivity of cumulative production to stimulated reservoir volume [31]

In some recent publications and presentations, the concept of Stimulated Reservoir Volume (SRV) has been linked to microseismic. In other words, it is advocated that by collecting and interpreting microseismic data and identifying microseismic events in a shale well that has been subject to multistage hydraulic fracturing, one can estimate the size of the Stimulated Reservoir Volume. As we will show in the next section, it should be noted that the evidences supporting such claims are equally countered by evidences that dispute them. Furthermore, it has been shown that misinterpreting the size of the Stimulated Reservoir Volume can result in large discrepancies in forecasting the potentials of a given well (see Fig. 2.11). It is a well-established concept that productions from shale wells to a large degree are a function of the amount and the extent of contact that is made with the rock. Therefore, the notion of production being very sensitive to estimation of the size and conductivity of the Stimulated Reservoir Volume is logically sound.

The sensitivity of production from shale wells to the size and the conductivity assigned to the Stimulated Reservoir Volume explains the uncertainties associated with the forecasts that are made using this technique. Although there have been attempts to address the dynamic nature of the SRV by incorporating stress dependent permeability (opening and closure of the fractures as a function of time and production), the entire concept remains in the realm of creative adaptation of existing tools and techniques to solve a new problems. In the opinion of the author, while SRV serves the purposes of modeling and history matching the observed production from a well, its contribution to forecasting the production (looking forward) is questionable, at best. Furthermore, SRV techniques are incapable of making serious contribution to designing an optimum frac job specific to a given well (looking backward).

2.10   Microseismic

The utility of microseismic events (as it is interpreted today from the raw data) to estimate Stimulated Reservoir Volume is at best inconclusive. While it has been shown that microseismic may provide some valuable information regarding the effectiveness of the hydraulic fractures in Eagle Ford Shale [32], the lack of correlation between recorded and interpreted microseismic data and the results of production logs in Marcellus Shale has been documented [33]. In some shale reservoirs, such as Marcellus, as shown in Fig. 2.12, although the current interpretation of microseismic raw data shows locations in the reservoir where "something" is happening or has happened, it does not seem to have much to do with the most important parameter that all parties are interested in, i.e., the production. The proven and independently verified value-added by microseismic as a tool for assessing the effectiveness of hydraulic fracture in production is a debatable issue that remains to be settled as more data becomes available and is published. Therefore, using the extent of microseismic events as an indicator for Stimulated Reservoir Volume seem to be a premature conclusion that has more to do with forceful justification of the utilization of the data that has cost a lot of money to generate than actual utilization of such data.

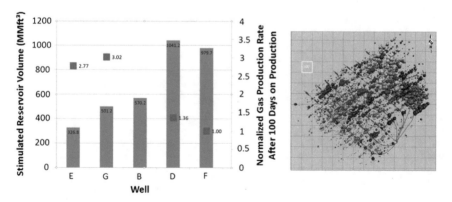

Fig. 2.12 Microseismic events, stimulated reservoir volume, and their contribution to production [33]

Due to its interpretive nature "Soft Data" cannot be used as optimization variables. In other words, one cannot expect to design a particular frac job that results in a well behaved induced fracture with a _designed_ half length, height, and conductivity by tweaking the amount of fluid and proppant that is injected. Similarly, _designing_ SRV (size and permeability) by modifying the amount of fluid and proppant that is injected during a frac job or by modifying the injection rate and pressure is not an option.[7] Therefore, although "Soft Data" may help engineers and modelers during the history matching process, it fails to provide a means for truly analyzing the impact of what is actually done during a frac job.

---

[7]Those who have opted to correlate "hard data" to Stimulated Reservoir Volume through microseismic events, are either technically too naïve to realize the premature nature of this effort, or trying to justify a service that is provided by their business partners.

Chapter 3
Shale Analytics

The realization that much value can be extracted from the data that is routinely collected (and much of it is left unused) during drilling, completion, stimulation, work over, injection, and production operations in the upstream exploration and production industry, have resulted in a growing interest in the application of data-driven analytics in our industry. Related activities that had been regarded as exotic academic endeavors have now come to the forefront and are attracting much attention. New startups are entering the marketplace, some with good products and expertise and others purely based on marketing gimmicks and opportunistic intuitions.

Petroleum Data Analytics (PDA) is the application of data-driven analytics and Big Data Analytics in the upstream oil and gas industry. It is the application of a combination of techniques that make the most of the data collected in the oil and gas industry in order to analyze, model, and optimize the production operations. Since the point of departure for this technology is data rather than physics and geology, it provides an alternative to conventional solutions that have been used in the industry for the past century. Shale Analytics is the implementation of Petroleum Data Analytics in all shale related issues. Since our understanding of the physics and the mechanics of the storage and transport phenomena in shale is quite limited, Shale Analytics can play a vital role in extracting much needed value from the data that is collected during the development of shale assets.

Shale Analytics is defined as the application of Big Data Analytics (data science, including data mining, artificial intelligence, machine learning and pattern recognition) in shale. Shale Analytics encompasses any and all data-driven techniques, workflows, and solutions that attempt to increase recovery and production efficiency from shale plays. Unlike conventional techniques such as Rate Transient Analysis (RTA), and numerical simulation that are heavily dependent on soft data such as fracture half-length, fracture height, fracture width, and fracture conductivity, Shale Analytics concentrates on using hard data (field measurements) in order to accomplish all its tasks that include but are not limited to:

1. Detail examination of the historical completion practices implemented on wells that are already producing (our experience shows that given the very large number of wells that have been drilled, completed, and are being produced, in the past several years, the perception of what has been done does not usually match the reality),
2. Finding trends and patterns in the seemingly chaotic behavior of the parameters that have been measured or used for design,
3. Identifying the importance of each reservoir and design parameter and finding the main drivers that are controlling the production,
4. Classifying and ranking areas in the field that may respond similarly to certain types of completion designs,
5. Building models with predictive capabilities that can calculate (estimate) well performance (production) based on measured reservoir characteristics, well spacing, completion parameters, and detail frac job practices,
6. Validating the predictive models with blind wells (wells set aside from the start and never used during the development of the predictive model),
7. Generating well-behaved type curves for different areas of the field that are capable of summarizing well performance as a function of multiple reservoir characteristics and design parameters,
8. Combining the predictive models with Monte-Carlo Simulation in order to be able to:

   a. Quantify the uncertainties associated with well productivity,
   b. Measure and compare, the quality of the historical frac jobs performed in the field,
   c. Determine the amount of reserve and production that have potentially been missed due to the sub-optimal completion practices,
   d. Measure and rank the accomplishments of the service companies in design and implementation of the completions,
   e. Rank the degree of success of the previous completions and stimulation practices,

9. Combining the predictive model with evolutionary optimization algorithms in order to identify the optimum (or near-optimum) frac designs for new wells.

10. Mapping the natural fracture network as a function of all the field measurements.
11. Identify and rank re-frac candidate wells, and recommend most appropriate completion design.

Shale Analytics has demonstrated its capabilities to accomplish the tasks enumerated above for more than 3000 wells in Marcellus, Utica, Eagle Ford, Bakken, and Niobrara shale. The success of Shale Analytics is highly dependent on close integration of domain expertise (practical knowledge of geology, petrophysics, and geophysics, as well as reservoir and production engineering) with state of the art in machine learning, artificial intelligence, pattern recognition, and data mining that combine both supervised and un-supervised data-driven algorithms.

Interest in Big Data Analytics is on the rise in our industry. Most of the operators have been active in forming data science and data analytics divisions. Even at the time that many drilling, reservoir, and production engineering jobs are at risk, operators and service companies are hiring data scientists. However, in the author's opinion, some companies are not taking the best route to take maximum advantage of what Big Data Analytics has to offer. The management must realize that if Big Data Analytics is not delivering tangible results in their operations and if data science is not fulfilling the promises made during the hype, the problem may be in the approach implemented to incorporate Big Data Analytics in the company. Of course, in order not to make themselves look bad, many decision makers are not ready to openly admit the impotent of the implemented approaches, but the final results in many companies is too telling to be ignored. Following paragraphs present the author's view on why the current approach in implementing Big Data Analytics and Data Science in our industry is facing obstacles and has been less than optimal, while it is flourishing in other industries.

Since its introduction as a discipline in mid-90s "Data Science" has been used as a synonym for applied statistics. Today, Data Science is used in multiple disciplines and is enjoying immense popularity. What has been causing confusion is the essence of Data Science as it is applied to physics-based disciplines such as oil and gas industry versus non-physics-based disciplines. Such distinctions surface once Data Science is applied to industrial applications and when it starts moving above and beyond simple academic problems.

So what is the difference between Data Science as it is applied to physics-based versus non-physic-based disciplines? When Data Science is applied to non-physics-based problems, it is merely applied statistics. Application of Data Science in social networks and social media, consumer relations, demographics, or politics (some may even include medical and/or pharmaceutical sciences to this list) takes a purely statistical form, since there are no sets of governing partial differential (or other mathematical) equations that have been developed to model human behavior or to the respond of human physiology to drugs. In such cases (non-physics-based areas), relationship between correlation and causation cannot be resolved using physical experiments and usually, as long as they are not absurd, are justified or explained, by scientist and statisticians, using psychological, sociological, or biological reasoning.

On the other hand, when Data Science is applied to physics-based problems such as self-driving cars, multi-phase fluid flow in reactors (CFD), or in porous media (reservoir simulation and modeling), and completion design and optimization in shale (Shale Analytics), it is a completely different story. The interaction between parameters that is of interest to physics-based problem solving, despite their complex nature, have been understood and modeled by scientists and engineers for decades. Therefore, treating the data that is generated from such phenomena (regardless whether it is measurements by sensors or generated by simulation) as just numbers that need to be processed in order to learn their interactions, is a gross mistreatment and over-simplification of the problem, and hardly ever generates useful results. That is why many of such attempts have, at best, resulted in unattractive and mediocre outcomes. So much so that many engineers (and scientists) have concluded that Data Science has little serious applications in industrial and engineering disciplines.

The question may rise that if the interaction between parameters that is of interest to engineers and scientists have been understood and modeled for decades, then how could Data Science contribute to industrial and engineering problems? The answer is: "considerable (and sometimes game changing and transformational) increase in the efficiency of the problem solving". So much so that it may change a solution from an academic exercise into a real-life solution. For example, many of the governing equations that can be solved to build and control a driverless car are well known. However, solving these complex set of high order, non-linear, partial differential equations and incorporating them into a real-time and real-life process that actually controls and drives a car in the street, is beyond the capabilities of any computer today (or in the foreseeable future). Data driven analytics and machine learning contribute significantly to accomplishing such tasks.There is a flourishing future for Data Science as the new generation of engineers and scientists are exposed to, and start using it in their everyday life. The solution is (a) to clarify and distinguish the application of Data Science to physics-based versus non-physics-based disciplines, (b) to demonstrate the useful and game changing applications of Data Science in engineering and industrial applications, and (c) to develop a new generation of engineers and scientists that are well versed in Data Science. In other words, the objective should be to train and develop engineers that understand and are capable of efficiently applying Data Scientist to problem solving. The techniques that are incorporated in author's work to perform the type of analysis, modeling, and optimization that is presented in this book include artificial intelligence and data mining. More specifically, artificial neural networks, evolutionary optimization, fuzzy set theory, as well as supervised and unsupervised hard and soft cluster analysis. Following sections in this chapter provide some brief explanations on each of these techniques. Much more can be learned about these technologies by digging deeper into the references that are provided.

3.1 Artificial Intelligence

Artificial Intelligence (AI) is the area of computer science focusing on creating machines that can engage on behaviors that humans consider intelligent. The ability to create intelligent machines has intrigued humans since ancient times and today with the advent of the computer and 50 years of research into AI programming techniques, the dream of smart machines is becoming a reality. Researchers are creating systems which can mimic human thoughts, understand speech, beat the best human chess player, win challenging "Jeopardy" contests, and countless other feats never before possible.

Artificial intelligence is a combination of computer science, physiology, and philosophy. AI is a broad topic, consisting of different fields, from machine vision to expert systems. The element that the fields of AI have in common is the creation of machines that can "think." In order to classify machines as "thinking," it is necessary to define intelligence. To what degree does intelligence consist of, for example, solving complex problems, or making generalizations and relationships? And what about perception and comprehension?

Artificial intelligence may be defined as a collection of analytic tools that attempts to imitate life. Artificial Intelligence techniques exhibit an ability to learn and deal with new situations. Artificial neural networks, evolutionary programming, and fuzzy logic are among the technologies that are classified as AI. These techniques possess one or more attributes of "reason," such as generalization, discovery, association, and abstraction. In the last two decade AI has matured to a set of analytic tools that facilitate solving problems that were previously difficult or impossible to solve. The trend now is the integration of these tools, as well as with conventional tools such as statistical analysis, to build sophisticated systems that can solve challenging problems. These tools are now used in many different disciplines and have found their way into commercial products. AI is used in areas such as medical diagnosis, credit card fraud detection, bank loan approval, smart household appliances, automated subway system controls, self-driving cars, automatic transmissions, financial portfolio management, robot navigation systems, and many more.

In the oil and gas industry these tools have been used to solve problems related to drilling, reservoir simulation and modeling, pressure transient analysis, well log interpretation, reservoir characterization, and candidate well selection for stimulation, among other things.

3.2 Data Mining

As the volume of data increases, human cognition is no longer capable of deciphering important information from it by conventional techniques. Data mining and machine learning techniques must be used in order to deduce information and knowledge from the raw data that resides in the databases.

Data mining is the process of extracting hidden patterns from data. With a marked increase in the amount of data that is being routinely collected, data mining is becoming an increasingly important tool to transform the collected data into information. Although the incorporation of Data Mining in the exploration and production industry is relatively new, it has been commonly used in a wide range of applications, such as marketing, fraud detection, and scientific discovery. Data Mining can be applied to datasets of any size. However, while it can be used to uncover hidden patterns in data that has been collected, obviously it can neither uncover patterns which are not already present in the data, nor can it uncover patterns in data that has not been collected.

Data Mining (sometimes referred to as Knowledge Discovery in Databases—KDD) has been defined as "The nontrivial extraction of implicit, previously unknown, and potentially useful information from data." It uses AI, machine learning, statistical, and visualization techniques to discover and present knowledge in a form which is easily comprehensible to humans.

Data Mining has been able to grasp the attention of many in the field of scientific research, businesses, banking sector, intelligence agencies, and many others from the early days of its inception. However, its use was not as easy as it is now. Data Mining is used by businesses to improve marketing and to understand the buying patterns of clients. Attrition Analysis, Customer Segmentation, and Cross Selling are the most important ways through which Data Mining is showing new ways in which businesses can increase revenue.

Data Mining is used in the banking sector for credit card fraud detection by identifying the patterns involved in fraudulent transactions. It is also used to reduce credit risk by classifying a potential client and predicting bad loans. Data Mining is used by intelligence agencies like FBI and CIA to identify threats of terrorism. After the 9/11 incident Data Mining has become one of the prime means to uncover terrorist plots.

In the popular article, "IT Does Not Matter" [35] Nicholas Carr argued that the use of IT is nowadays so widespread that any particular organization does not have any strategic advantage over the others due to the use of IT. He concludes that IT has lost its strategic importance. However, it is the view of the author that in today's E&P landscape Data Mining has become the sort of tool that can provide a strategic and competitive advantage for those that have the foresight to embrace it in their day to day operations. It is becoming more and more evident that NOCs, IOCs, and independent operators can create strategic advantages over theirs competitors by making use of Data Mining to get important insights from the collected data.

3.2.1 Steps Involved in Data Mining

There are various steps that are involved in mining data. Following are included among these steps:

(a) **Data Integration:** The reality is that in today's oil and gas industry data is never in the form that one needs in order to perform data mining. Usually there are multiple sources for data and data exists in several databases. Data needs to be collected and integrated from different sources in order to be prepared for data mining.

(b) **Data Selection:** Once the data is integrated it usually is used for specific purpose. The collected data needs to be studied and the specific parts of the data that lends itself to the task at hand in an organization should be selected for the given data mining project. For example, HR data may not be needed for a drilling project.

(c) **Data Cleansing:** The data that has been collected is usually not clean and may contain errors, missing values, noise, or inconsistencies. Different techniques need to be applied to the selected data in order to get rid of such anomalies.

(d) **Data Abstraction and Summarization:** The data that has been collected especially if it is operational data may need to be summarized while main essence of its behavior remains intact. For example, if pressure and temperature data is collected via a permanent down-hole gauge at one second frequency, it may need to be summarized or abstracted into minute data before use in a specific data mining project.

(e) **Data Transformation:** Even after cleaning and abstraction data may not be ready for mining. Data usually needs to be transformed into forms appropriate for mining. The techniques used to accomplish this are smoothing, aggregation, normalization, etc.

(f) **Data Mining:** Machine learning and techniques such as clustering and association analysis are among many different techniques used for data mining. Both descriptive and predictive data mining may be applied to the data. The objectives of the project will determine the type and the techniques used in data mining.

(g) **Pattern Evaluation and Knowledge Presentation:** This step involves visualization, transformation, and removing redundant patterns from the patterns generated.

(h) **Decisions/Use of Discovered Knowledge:** Use of the knowledge acquired to make better decisions is the essence of this last step.

3.3 Artificial Neural Networks

Much has been written about artificial neural networks. There are books and articles that can be accessed for deep understanding of the topic and all relevant algorithms. The objective here is to provide a brief overview, sufficient enough to make the understanding of the topics presented in this book, easy to follow. For more detailed understanding of this technology, it is highly recommended that the reader refer to the books and articles that have been referenced here.

Neural network research can be traced back to a paper by McCulloch and Pitts [34]. In 1958 Frank Rosenblatt invented the Perceptron [35]. Rosenblatt proved that given linearly separable classes, a perceptron would, in a finite number of training trials, develop a weight vector that will separate the classes (a pattern classification task). He also showed that his proof holds independent of the starting value of the weights. Around the same time Widrow and Hoff [36] developed a similar network called Adeline. Minskey and Papert [37] in a book called "Perceptrons" pointed out that the theorem obviously applies to those problems that the structure is capable of computing. They showed that elementary calculations such as simple "exclusive or" (XOR) problems cannot be solved by single layer Perceptrons.

Rosenblatt [35] had also studied structures with more layers and believed that they could overcome the limitations of simple Perceptrons. However, there was no learning algorithm known which could determine the weights necessary to implement a given calculation. Minskey and Papert doubted that one could be found and recommended that other approaches to AI should be pursued. Following this discussion, most of the computer science community left the neural network paradigm for twenty years [38]. In early 1980s, Hopfield was able to revive the neural network research. Hopfield's efforts coincided with development of new learning algorithms such as back-propagation. The growth of neural network research and applications has been phenomenal since this revival.

3.3.1   Structure of a Neural Network

An artificial neural network is an information processing system that has certain performance characteristics in common with biological neural networks. Therefore, it is appropriate to briefly describe a biological neural network before offering a detail definition of artificial neural networks.

All living organisms are made up of cells. The basic building blocks of the nervous system are nerve cells, called neurons. Figure 3.1 shows a schematic diagram of two bipolar neurons. A typical neuron contains a cell body, where the nucleus is located, dendrites, and an axon. Information in the form of a train of electro-chemical pulses (signals) enters the cell body from the dendrites. Based on the nature of this input the neuron will activate in an excitatory or inhibitory fashion and provides an output that will travel through the axon and connects to other neurons where it becomes the input to the receiving neuron. The point between two neurons in a neural pathway, where the termination of the axon of one neuron comes into close proximity with the cell body or dendrites of another, is called a synapse. The signals traveling from the first neuron initiate a train of electro-chemical pulse (signals) in the second neuron.

It is estimated that the human brain contains on the order of 10–500 billion neurons [39]. These neurons are divided into modules and each module contains about 500 neural networks [40]. Each network may contain about 100,000 neurons in which each neuron is connected to hundreds to thousands of other neurons. This

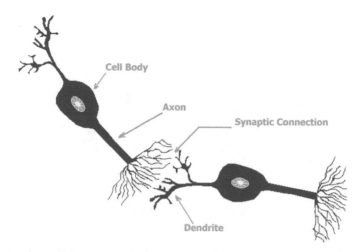

Fig. 3.1 Schematic diagram of two bipolar neurons

architecture is the main driving force behind the complex behavior that comes so natural to us. Simple tasks such as catching a ball, drinking a glass of water, or walking in a crowded market require so many complex and coordinated calculations that sophisticated computers are unable to undertake the task, and yet is done routinely by humans without a moment of thought.

This becomes even more interesting when one realizes that neurons in the human brain have cycle time of about 10–100 ms while the cycle time of a typical desktop computer chip is measured in nanoseconds (about 10 million times faster than human brain). The human brain, although million times slower than common desktop PCs, can perform many tasks orders of magnitude faster than computers because of it massively parallel architecture.

Artificial neural networks are a rough approximation and simplified simulation of the process explained above. An artificial neural network can be defined as an information processing system that has certain performance characteristics similar to biological neural networks. They have been developed as generalization of mathematical models of human cognition or neural biology, based on the assumptions that:

1. Information processing occurs in simple processing elements, called neurons.
2. Signals are passed between neurons over connection links.
3. Each connection link has an associated weight, which, in a typical neural network, multiplies the signal being transmitted.
4. Each neuron applies an activation function (usually nonlinear) to its net input to determine its output signal [41].

Figure 3.2 is a schematic diagram of a typical neuron (processing element) in an artificial neural network. Output from other neurons is multiplied by the weight of the connection and enters the neuron as input. Therefore, an artificial neuron has

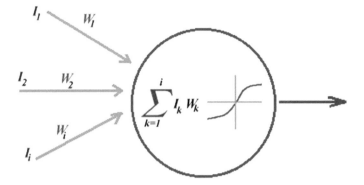

Fig. 3.2 Schematic diagram of an artificial neuron or a processing element

many inputs and only one output. The inputs are summed and subsequently applied to the activation function and the result is the output of the neuron.

3.3.2 Mechanics of Neural Networks Operation

An artificial neural network is a collection of neurons that are arranged in specific formations. Neurons are grouped into layers. In a multilayer network there are usually an input layer, one or more hidden layers, and an output layer. The number of neurons in the input layer corresponds to the number of parameters that are being presented to the network as input. The same is true for the output layer. It should be noted that neural network analysis is not limited to a single output and that neural networks can be trained to build data-driven models with multiple outputs. The neurons in the hidden layer or layers are mainly responsible for feature extraction.

They provide increased dimensionality and accommodate tasks such as classification and pattern recognition. Figure 3.3 is a schematic diagram of a fully connected three layered neural network. There are many kinds of neural networks.

Fig. 3.3 Schematic diagram
of a three-layer neuron
network

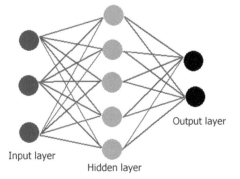

Neural network scientists and practitioners have provided different classifications for neural networks. One of the most popular classifications is based on the training methods. Neural networks can be divided into two major categories based on the training methods, namely, supervised and unsupervised neural networks. Unsupervised neural networks, also known as self-organizing maps, are mainly clustering and classification algorithms. They have been used in oil and gas industry to interpret well logs and to identify lithology. They are called unsupervised simply because no feedback is provided to the network by the person that is training the neural network. The network is asked to classify the input vectors into groups and clusters. This requires a certain degree of redundancy in the input data and hence the notion that redundancy is knowledge [42].

Most of the practical and useful neural network applications in the upstream oil and gas industry are based on supervised training algorithms. During a supervised training process both input and output are presented to the network to permit learning on a feedback basis. A specific architecture, topology, and training algorithm are selected and the network is trained until it converges to an acceptable solution. During the training process neural network tries to converge to an internal representation of the system behavior. Although by definition neural networks are model-free function approximators, some people choose to call the trained network a neuro-model. In this book our preferred terminology is "data-driven" model.

The connections correspond roughly to the axons and synapses in a biological system, and they provide a signal transmission pathway between the nodes. Several layers can be interconnected. The layer that receives the inputs is called the input layer. It typically performs no function other than the buffering of the input signal. In most cases, the calculation performed in this layer is a normalization of the input parameters so that parameters such as porosity (which usually is represented in fractions) and initial pressure (which is usually in thousands of psi) would be treated equally by the neural network at the start of the training process. The network outputs are generated from the output layer. Any other layers are called hidden layers because they are internal to the network and have no direct contact with the external environment. Sometimes they are likened to a "black box" within the network system. However, just because they are not immediately visible does not mean that one cannot examine the function of those layers. There may be zero to several hidden layers in a neural network. In a fully connected network every output from one layer is passed along to every node in the next layer.

In a typical neural data processing procedure, the database is divided into three separate portions called training, calibration, and verification sets. The training set is used to develop the desired network. In this process (depending on the training algorithm that is being used), the desired output in the training set is used to help the network adjust the weights between its neurons or processing elements. During the training process the question arises as when to stop the training? How many times should the network go through the data in the training set in order to learn the system behavior? When should the training stop? These are legitimate questions, since a network can be overtrained. In the neural network related literature over-training is also referred to as memorization. Once the network memorizes a dataset, it would be incapable of generalization. It will fit the training dataset quite accurately, but suffers in generalization. Performance of an overtrained neural network is similar to a complex nonlinear regression analysis.

Overtraining does not apply to some neural network algorithms simply because they are not trained using an iterative process. Memorization and overtraining is applicable to those networks that are historically among the most popular ones for engineering problem-solving. These include back-propagation networks that use an iterative process during the training.

In order to avoid overtraining or memorization, it is a common practice to stop the training process very often and apply the network to the calibration dataset. Since the output of the calibration dataset is not presented to the network (during the training), one can evaluate network's generalization capabilities by how well it predicts the calibration set's output. Once the training process is completed successfully, the network is applied to the verification dataset.

During the training process each artificial neuron (processing element) handles several basic functions. First, it evaluates input signals and determines the strength of each one. Second, it calculates a total for the combined input signals and compares that total to some threshold level. Finally, it determines what the output should be. The transformation of the input to output—within a neuron—takes place using an activation function. Figure 3.4 shows two of the commonly used activation (transfer) functions.

All the inputs come into a processing element (in the hidden layer) simultaneously. In response, neuron either "fires" or "doesn't fire" depending on some threshold level. The neuron will be allowed a single output signal, just as in a biological neuron—many inputs, one output. In addition, just as things other than inputs affect real neurons, some networks provide a mechanism for other influences. Sometimes this extra input is called a bias term, or a forcing term. It could also be a forgetting term, when a system needs to unlearn something [43].

Initially each input is assigned a random relative weight (in some advanced applications—based on the experience of the practitioner—the relative weight assigned initially may not be random). During the training process the weights of the inputs are adjusted. The weight of the input represents the strength of its connection to the neuron in the next layer. The weight of the connection will affect

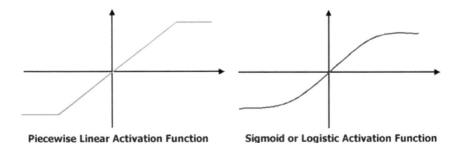

Piecewise Linear Activation Function   **Sigmoid or Logistic Activation Function**

Fig. 3.4 Commonly used activation functions in artificial neurons

the impact and the influence of that input. This is similar to the varying synaptic strengths of biological neurons. Some inputs are more important than others in the way they combine to produce an impact. Weights are adaptive coefficients within the network that determine the intensity of the input signal. The initial weight for a processing element could be modified in response to various inputs and according to the network's own rules for modification.

Mathematically, we could look at the inputs and the weights on the inputs as vectors, such as I_1, I_2, I_3, I_4 ... I_n for inputs and as W_1, W_2, W_3, W_4 ... W_n for weights. The total input signal is the dot, or inner, product of the two vectors. Geometrically, the inner product of two vectors can be considered as a measure of their similarity. The inner product is at its maximum if the vectors point in the same direction. If the vectors point in opposite directions (180°), their inner product is at its minimum.

Signals coming into a neuron can be positive (excitatory) or negative (inhibitory). A positive input promotes the firing of the processing element, whereas a negative input tends to keep the processing element from firing. During the training process some local memory can be attached to the processing element to store the results (weights) of previous computations. Training is accomplished by modification of the weights on a continuous basis until convergence is reached. The ability to change the weights allows the network to modify its behavior in response to its inputs, or to learn. For example, suppose a network being trained to correctly calculate the initial production from a newly drilled well. At the early stages of the training the neural network calculates initial production from the new well to be 150 bbls/day whereas the actual initial production is 1000 bbls/day. On successive iterations (training), connection weights that respond to an increase in initial production (the output of the neural network) are strengthened and those that respond to a decrease, are weakened until they fall below the threshold level and the correct calculation of the initial production is achieved.

In the back-propagation algorithm [44] (one of the most commonly used supervised training algorithms in the upstream oil and gas operations) the network output is compared with the desired output—which is part of the training dataset, and the difference (error) is propagated backward through the network. During this back-propagation of error the weights of the connections between neurons are adjusted. This process is continued in an iterative manner. The network converges when its output is within acceptable proximity of the desired output.

3.3.3   Practical Considerations During the Training of a Neural Network

There are substantial amounts of art involved in the training of a neural network. In this section of the book the objective is to share some personal experiences that have been gained throughout many years of developing data-driven models for the oil and gas related problems. These are mainly practical considerations and may or may not agree with similar practices in other industries, but they have worked very well for the author in the past two decades.

Understanding how machines learn is not complicated. The mathematics associated with neural networks including how they are built and trained, is not really complex [44]. It includes vector calculus and differentiation of some normal functions. While being a good reservoir or production engineer and a capable petroleum engineering modeler is essential for building and effectively utilizing the data-driven models that are covered in this book. In order to be a capable Petroleum Data Scientist, while it is necessary to be a competent petroleum Engineer, it is not a requirement to have degrees in mathematics, statistics, or machine learning.

For developing a functional data-driven models there are no need to be an expert in machine learning or artificial neural networks. What is required though is to be able to understand the fundamentals of this technology and eventually become an effective user and practitioner of the technology. Although such skills are not taught as part of any petroleum engineering curriculum at the universities, acquiring such skills is not a far reaching task and any petroleum engineer with a bachelor degree should be able to master it with some training and effort. It is important to understand and subscribe to the philosophy of machine learning. This means that although as an engineer you have learned to solve problems in a particular way, you need to understand and accept that there are more than one way to solve engineering related problems.

The technique that engineers have been using to solve problem follows a well-defined path of identifying the parameters involved and then constructing the relationships between the parameters (using mathematics) to build a model. The philosophy of solving a problem using machine learning is completely different. Given the fact that in machine learning algorithms such as artificial neural networks, models are built using data, the path to follow in order to solve an engineering problem changes from that you have learned as a petroleum engineer. To solve problems using data, you have to be able to teach a computer algorithm, about the problem and its solutions. This process is called supervised learning. You have to create (actually collect) a large number of records (examples) that include the inputs (parameters involved) and the outputs (the solution you are trying to solve for). During training, these records (the coupled input–output pairs) are presented to the machine learning algorithm and by repetition (and some learning algorithms and repetitions) the machine will eventually learn how the problem is solved. The algorithm does this by building an internal representation of the mapping between inputs and outputs.

In the previous section of this book fundamentals of this technology was covered. In this section some practical aspects of this technology will be briefly discussed. These practical aspects will help a data-driven modeler learn how to train good and useful neural networks. The neural network training includes several steps that are covered in this section.

3.3.3.1 Selection of Input Parameters

Since all models are wrong the scientist cannot obtain a "correct" one by excessive elaboration. On the contrary, following William of Occam,[1] the scientist should seek an economical description of natural phenomena. Just as the ability to devise simple but evocative models is the signature of the great scientist, so overelaboration and over-parameterization are often the mark of mediocrity [45]. In order not to over-parameterize our neural network model we need to use the right number of variables.

Selection of the input parameters that are going to be used to train a neural network from amongst the variables (potential inputs) that have been assimilated in the database is not a trivial procedure. The database generated for a given project usually includes a very large number of parameters all of which are potential input parameters to the neural networks that are going to be trained for the data-driven model. They include static parameters, and dynamic parameters as well as similar parameters for several offset wells. These parameters are designated as columns in a flat file that is eventually used to train data-driven models (neural networks).

Not all of the parameters that are included in the database are used to train the neural networks. Actually, it is highly recommended that the number of parameters that are used to build (train, calibrate and validate) the neural network be limited. This limitation of the input parameters should not be interpreted that other parameters do not play any role in forming or calculating the output of the model. Elimination of some of the parameters and use of some others to build the data-driven model simply means that a subset of the parameters play such an important role in the determination of the model output that they overshadow (or sometimes implicitly represent) the impact of other parameters. Such that a model can be developed using only these parameters and safely ignoring others.

Therefore, only a subset of these parameters should be selected and used as the input to the data-driven models. Experience with developing successful data-driven models has shown that the process of selecting the parameters that must be used as input to the model needs to satisfy the following three criteria:

1. The impact (influence) of all the exiting parameters (in the database) on the model output should be identified and ranked. Then the top "x" percent of these ranked parameters should be used as input in the model. This is easier said than done. There are many techniques that can be used to help data-driven modeler in identifying the influence of parameters on a selected output. These techniques can be as simple as linear regression and as complex as Fuzzy Pattern Recognition.[2] Some have used Principal Component Analysis [46] to accomplish this task.

---

[1]Occam's razor is a problem-solving principle devised by William of Ockham (c. 1287–1347). The principle states that among competing hypotheses, the one with the fewest assumptions should be selected. In the absence of certainty, the fewer assumptions that are made, the better.

[2]This is a proprietary algorithm developed by Intelligent Solutions, Inc. and used in their Data-driven Modeling software application called IMprove™ (www.IntelligentSolutionsInc.com).

2. In the list of input parameters that are identified to be used in the training of the neural network, there must exist parameters that can validate the physics and/or the geology of the model. If such parameters are already among the highly ranked parameters in the previous step, then great, otherwise, the data-driven modeler must see to it that they are included in the model. Being able to verify that the data-driven model has understood the physics and honors it, is an important part of data-driven modeling.

3. In many cases the data-driven model is developed in order to optimize production. Identification of optimized choke setting during production is a good example of such a situation. In such cases, parameters that are needed in order to optimize production should be included in the set of input parameters. If the optimization parameters are already among the highly ranked parameters in the previous step, then great, otherwise, the data-driven modeler must see to it that they are included in the model.

Machine learning literature includes many techniques for this purpose. In these literatures the technology is referred to as "Feature Selection".

3.3.3.2 Partitioning the Dataset

Data in the spatio-temporal database is transferred into a flat file once a superset of the parameters are selected to be used in the training of the neural networks. The data in the flat file needs to be partitioned into three segments: training, calibration, and validation. As it will be discussed in the next section, the way these segments are treated, determines the essence of the training process. In this section the characteristics of each of these data segments and their use and purpose is briefly discussed.

In general the largest of the three segments is the training dataset. This is the data that is used to train the neural network and create the relationships between the input parameters and the output parameter. Everything that one wishes to teach a data-driven model must be included in the training dataset. One must realize that the range of the parameters as they appear in the training set determines the range of the applicability of the data-driven model. For example, if the range of permeability in the training set is between 2 and 200 mD one should not expect the data-driven model to perform reasonably well for a data record with permeability values less than 2 mD and higher than 200 mD. This is due to the well-known fact that most machine learning algorithms, neural networks included, demonstrate great interpolative capabilities, even if the relationship between the inputs parameters and the output(s) is highly nonlinear. However, machine learning algorithms are not known for their extrapolative capabilities.

As we mentioned in Sect. 3.3.1, the input parameters are connected to the output parameter through a set of hidden neurons. The strength of the connections between neurons (between input neurons and hidden neurons, between hidden neurons with one another if such connections exists, and between hidden neurons and output neurons) are determined by the weight associated with each connection. During the training process the optimum weight of each connection is determined through an iterative learning process.

During the training process the weights between the neurons (also known as synaptic connections) in neural networks find their optimum value. The collection of these optimum values forms the coefficient matrices that are used to calculate the output parameter. Therefore, the role of the training dataset is to help the modeler to determine the strength between the neurons in a neural network. Convergence of a network to a desirable set of weights that will translate to a well-trained and smart neural network model depends on the information content of the training dataset. When the neural networks are being trained for data-driven model purposes, the size of the training dataset may be as high as 80 % or as low as 40 % of the entire dataset. This percentage is a function of the number of records in the database.

The calibration dataset is not used directly during the training process and it actually plays no direct role in changing the weights of the connections between the neurons. The calibration dataset is a blind dataset that is used after every epoch of training[3] in order to test the quality and the goodness of the trained neural network. In many circles this is also called the "Test" set. The calibration dataset is essentially a watch dog that observes the training process and decides when to stop the training process, since the network is only as good as its prediction of the calibration dataset (a randomly selected dataset that is actually a blind dataset).

Therefore, after every epoch of training (when the network get to see all the records in the training dataset, once) the weights are saved and the network is tested against the calibration dataset to see if there has been any improvement of network predictive performance against this blind dataset. This test of network predictive capabilities is performed after every epoch to monitor its generalization capabilities. Usually one or more metrics such as the R^2, the correlation coefficient, or the Mean Square Error (MSE) are used to calculate network's generalization capabilities. These metrics are used to determine how closely the set of synaptic connection weights will enable the calculation of the outputs as a function of the input parameters by comparing the output values computed by the neural networks against those measures in the field (the actual or real outputs) and used to train the neural network.

As long as this metric is improving for the calibration dataset, it means that the training can continue and the network is still learning. When the neural networks are being trained for data-driven modeling purposes, the size of the calibration dataset is usually between 10 and 30 % of the entire dataset, depending on the size of the database.

The last, but arguable the most important, dataset (segment) is the validation or verification dataset. This dataset plays no role during the training or calibration of the neural network. It has been selected and put aside from the very beginning to be used as a blind dataset. It literally sits on the sidelines and does nothing until the

---

[3]An epoch of training is completed when all the data records in the training set have been passed through the neural network and the error between neural network output and the actual field measurements are calculated.

training process is over. This blind dataset validates the generalization capabilities of the trained neural network.

While having no role to play during the training and calibration of the neural network, this dataset validates the robustness of the predictive capabilities of the neural network. The data-driven model that will result from the neural networks that is being trained is as good as the outcome of the validation or verification dataset. When the neural networks are being trained for data-driven modeling purposes, the size of the validation (verification) dataset is usually between 10 and 30 % of the entire dataset, depending on the size of the database.

Since the database is being partitioned into three datasets, it is important to make sure that the information content of these datasets are comparable to one another. If they differ, and a lot of time they will, then it would be best that the training set have the largest, most comprehensive information content of the three datasets. This will ensure a healthy training behavior and will increase the possibility of training a good and robust neural network. Information content and its relationship with entropy within the context of the information theory is an interesting subject that those involved with data-driven analytics and machine learning should understand [47].

3.3.3.3 Structure and Topology

The structure and the topology of a neural network are determined by several factors, and hypothetically can have an infinite number of possible forms. However, almost all of them include a combination of factors such as the number of hidden layers, the number of hidden neurons in each hidden layer, combination of the activation functions, and the nature of the connections between neurons. In this section the objective is to briefly discuss some of the most popular structures and specifically those that have shown success when used in the development of data-driven models for the oil and gas related applications. In other words, the intention is not to turn this chapter of the book into a neural network tutorial, rather to present practices that have proven successful during the development of data-driven models in the past for the author and can be used as "Rule of Thumb" for those that will be entering the world of data-driven modeling.

As far as the connection between neurons is concerned, the structures that have been used most successfully in data-driven models are fully connected neural networks. In fully connected networks every input neuron is connected to every hidden neuron and also, every hidden neuron is connected to the output neuron, the network is called a fully connected network, as shown in Figs. 3.5, 3.6 and 3.7.

Figure 3.5 shows the most simple and also the most popular type of neural networks for the development of data-driven models that form the main engines of the data-driven model. This is a simple, three-layer, fully connected neural network. The three layers are the input layer, the hidden layer, and the output layer. Furthermore, while the number of output neurons in the output layer can be more than one, our experience with data-driven model has shown that except in some specific situations, a single output neuron in the output layer performs best.

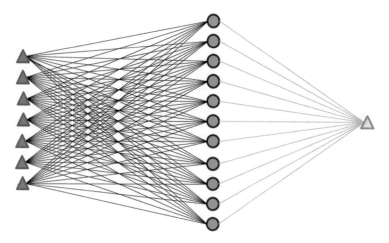

Fig. 3.5 A fully connected neural network with one hidden layer that includes 11 hidden neurons, seven input neurons and one output neuron

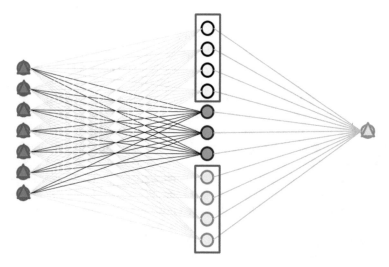

Fig. 3.6 A fully connected neural network with one hidden layer that includes three different sets of activation functions along its 11 hidden neurons, seven input neurons and one output neuron

Furthermore, after experiencing with a large number and variety of network structures, author's experience has shown that if you are not able to train a good network[4] using the simple structure as shown in Fig. 3.5, your chances of getting a

---

[4]What constitutes a "good network"? A good network is a network that can be trained and calibrated and validated. It can learn well and has robust predictive capabilities. The rest is very problem dependent.

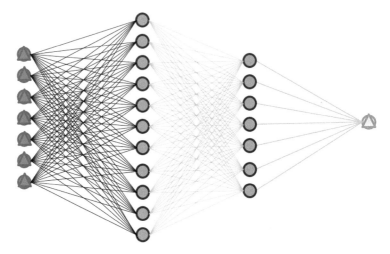

Fig. 3.7 A fully connected neural network with two hidden layer that includes 11 and seven hidden neurons, respectively, as well as seven input neurons and one output neuron

good neural network with any other structure will be slim. In other words, it is not the structure of the neural network that will make or break the success of your efforts in training a data-driven model, rather it is the quality and the information content of the database that determines your chances of being successful in developing a neural network, and with the same token, a data-driven model. If you see persistent issues in developing the data-driven model (the networks cannot be trained properly or do not have good predictive capabilities), you need to revisit your database rather than playing around with the structure and the topology of your neural network.

In the author's opinion, when the practitioners of the machine learning (specifically in the upstream oil and gas industry) concentrate their efforts, and naturally the presentation of their data-driven modeling efforts on the neural network structure and how it can be modified in order to control the training of the model, one must take that as an indication of naivety and lack of substance and skill of the practitioner in machine learning technology, and not an indication of expertise. You will find that an expert in the art and science of machine learning will concentrate mostly on the information content and details of the dataset being used to train the neural network as long as very basic issues (the number of hidden layers and hidden neurons, the learning rate and momentum, etc.) in the structure of the neural network has been reasonably determined to be solid. The details regarding the structure and the topology of a neural network can have enhancing impact on the results, but it will not make or break your data-driven model.

Sometimes, changing some of the activation functions can help in fine-tuning the neural network's performance as shown in Fig. 3.6. In this figure the hidden neurons are divided into three segments and each set of the hidden neurons can be assigned a different activation function. Some details about activation functions were covered in the previous sections. The author does not recommend the initial structure of a neural network model be designed as shown in Fig. 3.6, rather, if need be, the structure in Figs. 3.6 and 3.7 may be used to enhance the performance of a network that has shown serious promise to be a good network and its enhancement requires some fine-tuning.

Once the structure of the neural network has been determined, it is time to decide upon the learning algorithm. By far the most popular learning (training) algorithm is the "error back-propagation" [44] or simply referred to as "back-propagation." In this learning algorithm the network calculates a series of outputs based on the current values of the weights (strength of the connections between neurons—synaptic connections) and compares its calculated outputs for all the records with actual (measured) outputs (what it is trying to match).

The calculated error between the network output and the measured values (also known as target) are then back propagated throughout the network structure with the aim of modifying the synaptic weights between neurons as a function of the magnitude of the calculated error. This process is continued until the back-propagation of the error and modification of the connection weights no longer enhances the network performance.

Several parameters are involved and can be modified during this training process that can impact the progression of the network training. These parameters include network's learning rate and momentum for the weights between each set of neurons (layers) as well as the nature of the activation function. However, just like it was mentioned before, none of these factors will make or break a neural network, rather they can be instrumental in fine-tuning the result of a neural network. The information content of the database (which is essentially domain expertise related to petroleum engineering and geosciences) is the most important factor in the success or failure of a data-driven model.

3.3.3.4 The Training Process

Since neural networks are known to be universal function approximators, hypothetically speaking, they are capable of complete replication (reproduction) of the training dataset. In other words, given enough time and large enough number of hidden neurons, a neural network should be able to reproduce the outputs of the training set from all the inputs, with 100 % accuracy. This is something that one expects from a statistical approach or mathematical spline curve fitting process. Such a result form a neural network is highly *undesirable*, and must be avoided. This is due to the fact that a neural network that is so accurate on the training set, has literally memorized all the training records, and has next to no predictive value.

This is the process that is usually referred to as overtraining or overfitting and in the AI lingo is referred to as "memorization" and must be avoided. An overtrained neural network memorizes the data in the training set and can reproduce the output values, almost identically, and does not learn it. Therefore, it cannot generalize and will not be able to predict the outcome of new data records. Such a model (if it can actually be called a model) is merely a statistical curve fit, and has no value, whatsoever.

Some of the geostatistical techniques that are currently used by many of the commercial software applications to populate the geocellular models are examples of such technique. Some of the most popular geomodeling software applications that are quite cognizant of this fact, have incorporated neural networks as part of their tools. However, a closer look at the way neural networks have been implemented in these software applications reveals that they are merely a marketing gimmick and these software applications incorporate them as statistical curve fitting techniques, which make neural network as useless as other geostatistical techniques.

One of the roles of the calibration dataset is to prevent overtraining. It is a good practice to observe the network behavior during the training process in order to understand whether the neural network is in the process of converging to a solution or it needs attention of the modeler. There is much that can be learned from this observation. A simple plot of Mean Squared Error (MSE) versus number of training epochs displays the neural networks training and convergence behavior. Furthermore, if the MSE is plotted for both the training dataset and the calibration dataset (after every training epoch), much can be learned from their side-by-side behavior. Several examples of such plots are shown in Figs. 3.8 and 3.9.

Figures 3.8 and 3.9 include three sets of examples, each. Each example includes two graphs. The ones on the left, show the MSE versus Number of Epochs for the calibration dataset, and the ones on the right, show the MSE versus Number of Epochs for the training dataset. Please remember that training and calibration datasets are completely independent and have different sizes. Normally, the training dataset includes 80 % and the calibration dataset includes 10 % of the complete dataset. Each pair of graphs in Figs. 3.8 and 3.9 represent the training process of one neural networks. Each of these figures includes three examples.

During a healthy training process, the error in the calibration dataset (two dimensional graphs on the left side of Figs. 3.8 and 3.9) that is playing no role in changing the weights in a neural network is expected to behave quite similar to the error in the training dataset (two dimensional graphs on the right side of Figs. 3.8 and 3.9). Several example of a healthy training process is shown in Fig. 3.8. In the plots shown in Fig. 3.8 the MSE is plotted versus number of training epochs.

If the graphs in Fig. 3.8 are plots that are being updates in real-time, then the modeler can observe the error behavior of the training progress in real-time and decide whether the training process should continue or it should be stopped so that some modifications can to be made to the network structure or the datasets. Actions such as this may be necessary once it is decided that the training process has entered a potential dead-end (lack of convergence) and there will be no more learning.

A healthy training process is defined as one where continuous, effective learning is taking place and the network is getting better by each epoch of training. One of the indications of such healthy training process is the similarity of the behavior between the two plots, as shown in Fig. 3.8. In the three examples shown in this figure, errors in both calibration and training sets have similar slope and behavior. This is important since calibration dataset is blind and independent of the training

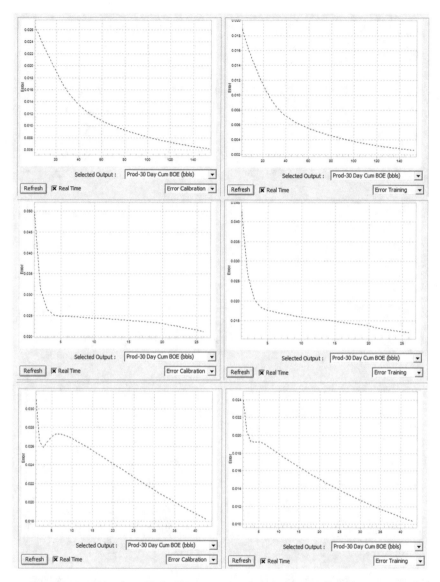

Fig. 3.8 Plot of Mean Square Error (MSE) as a function of number of training epochs. The plot on the *left* shows the error for the calibration dataset and the plot on the *right* shows the error for the training dataset. Three examples of a training process that is progressing in a satisfactory manner when the behavior of the errors mirrors one another

set and such similarities in the error behavior indicate an effective partitioning of the datasets.

On the other hand, an unhealthy training process is one that has the behavior of the error in the training and calibration datasets different and sometimes it starts

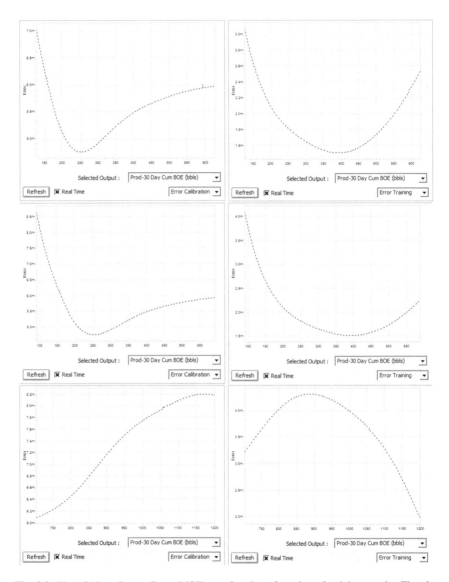

Fig. 3.9 Plot of Mean Square Error (MSE) as a function of number of training epochs. The plot on the *left* shows the error for the calibration dataset and the plot on the *right* shows the error for the training dataset. Three examples of a training process that is NOT progressing in a satisfactory manner when the behaviors of the errors are different with opposite slopes

moving in the opposite direction. Figure 3.9 shows three examples of unhealthy error behavior. This said, it also should be mentioned that given the nature of how gradient decent algorithms such as back-propagation works, it is expected that from time to time a difference in the error behavior between these two dataset be observed, but it would be temporary. In such cases, if you give the algorithm enough time, it will correct itself and the behavior of error start getting healthier. This is of course a function of the problem being solved and the prepared dataset that is being used and must be carefully observed and judged by the modeler.

At this point in time, a legitimate question that can be asked is: what would make a training behavior unhealthy and how can it be overcome? For example if the best network that is saved is the one with the best (highest) R^2 and/or lowest value of MSE for the calibration dataset, how can we try to avoid an early (premature) convergence? A premature convergence is defined as a situation where the error in the training dataset is decreasing while the opposite trend is observed in the error of calibration dataset as shown in Fig. 3.9. The answer to this question relies on the information content of the training, the calibration, and the validation datasets, as was mentioned in the previous section. In other words, one of the reasons such phenomena can take place is the way the database has been partitioned. To clarify this point an example is provided. The example is demonstrated using Figs. 3.10, 3.11 and 3.12.

In these figures you can see that the largest value of the field measurement (y axis) for the output (30 days cumulative production in BOE) in the training dataset is 3850 bbls (Fig. 3.10) while the larges value of the field measurement for the output in the validation dataset is 4550 bbls (Fig. 3.12). Clearly, the model is not being trained on any field measurements with values larger than 3850 bbls.

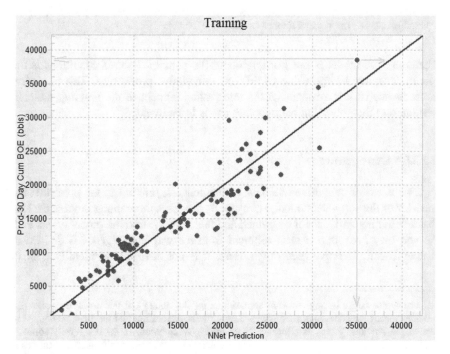

Fig. 3.10 Cross plot of observed (measured) output versus neural network prediction for the training dataset. The larges field measurement is 3850 bbls

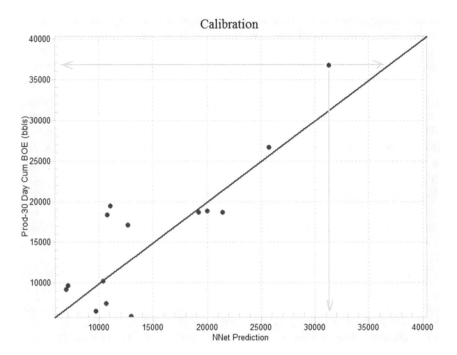

Fig. 3.11 Cross plot of observed (measured) output versus neural network prediction for the calibration dataset. The larges field measurement is 3700 bbls

Therefore, the model is not learning about the combined set of conditions is the reason for such a large value of 30 days cumulative production. This is a clear consequence of inconsistency of the information content of the training, the calibration, and the validation datasets that needs to be avoided.[5]

3.3.3.5 Convergence

In the context of data-driven models, convergence is referred to the point that the modeler or the software's intelligent agent that oversee the training process decides that a better network cannot be trained and therefore the training process must end. As you may note, this is a bit different from the way convergence is defined in mathematically iterative procedures. Here, it is not advisable to identify a small

---

[5]Author is not aware of any software applications for the training of the neural networks that provides means for addressing such issues. The software application that has been mentioned in one of the previous footnotes (IMprove™ by Intelligent Solutions, Inc.) is the only software application in the oil and gas industry that include means to detect and rectify such issues. This is due to identification and then addressing of the practical issues that can be encountered when data-driven models are built to address upstream exploration and production problems.

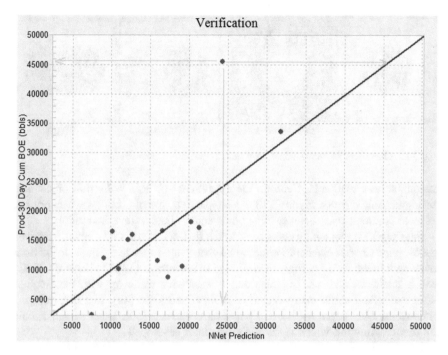

Fig. 3.12 Cross plot of observed (measured) output versus neural network prediction for the validation dataset. The larges field measurement is 4550 bbls

enough delta error for the convergence since such error value may never be achieved. The acceptable error in data-driven models is very much a problem dependent issue.

In data-driven models, the best type of convergence criteria is the highest R^2 or the lowest MSE for the calibration dataset. It is important to note that in many cases, these values can be a bit misleading (although they remain the best measure) and it is recommended to visually inspect the results for all the wells in the field, individually, before making a decision of whether to stop the training, or the search for a better data-driven model should still continue.

3.4 Fuzzy Logic

Todays' science is based on Aristotle's crisp logic formed more than two thousand years ago. The Aristotelian logic looks at the world in a bivalent manner, such as black and white, yes and no, and 0 and 1. Development of the set theory in the late 19th century by German mathematician George Cantor that was based on the Aristotle's bivalent logic made this logic accessible to modern science. Then, the subsequent superimposition of probability theory made the bivalent logic

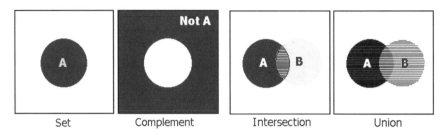

Fig. 3.13 Operations of conventional crisp sets

reasonable and workable. Cantor's theory defines sets as a collection of definite, distinguishable objects. Figure 3.13 is a simple example of Cantor's set theory and its most common operations such as complement, intersection, and union.

First work on vagueness dates back to the first decade of 1900, when American philosopher Charles Sanders Peirce noted that "vagueness is no more to be done away with in the world of logic than friction in mechanics [48]." In the early 1920s, Polish mathematician and logician Jan Lukasiewicz came up with three-valued logic and talked about many-valued or multivalued logic [49]. In 1937, quantum philosopher Max Black published a paper on vague sets [50]. These scientists built the foundation upon which fuzzy logic was later developed.

Lotfi A. Zadeh is known to be the father of fuzzy logic. In 1965, while he was the chair of the electrical engineering department at UC Berkeley, he published his landmark paper "Fuzzy Sets [51]" Zadeh developed many key concepts including the membership values and provided a comprehensive framework to apply the theory to many engineering and scientific problems. This framework included the classical operations for fuzzy sets, which comprises all the mathematical tools necessary to apply the fuzzy set theory to real world problems.

Zadeh used the term "fuzzy" for the first time, and with that he provoked many oppositions. He became a tireless spokesperson for the field. He was often harshly criticized. For example, Professor R.E. Kalman said in a 1972 conference in Bordeaux, "Fuzzification is a kind of scientific permissiveness; it tends to result in socially appealing slogans unaccompanied by the discipline of hard scientific work [52]." It should be noted that Kalman is a former student of Zadeh's and the inventor of the famous Kalman filter, a major statistical tool in electrical engineering that has also been used in computer assisted history matching in the oil and gas industry. Despite all the adversities fuzzy logic continued to flourish and has become a major force behind many advances in intelligent systems.

The term "fuzzy" carries a negative connotation in the western culture. The term "fuzzy logic" seems to both misdirect the attention and to celebrate mental fog [53]. On the other hand, eastern culture embraces the concept of coexistence of contradictions as it appears in the Yin–Yang symbol. While Aristotelian logic preaches "A," *or* "Not-A," Buddhism is all about "A," **and** "Not-A" (Fig. 3.14).

Many believe that the tolerance of eastern culture for such ideas was the main reason behind the success of fuzzy logic in Japan. While fuzzy logic was being

Fig. 3.14 The Yin–Yang
symbol

attacked in the United States, Japanese industries were busy building a multibillion
dollar industry around it. By the late 1990s, Japanese held more than 2000 fuzzy
related patents. They have used the fuzzy technology to build intelligent household
appliances such as washing machines and vacuum cleaners (Matsushita and
Hitachi), rice cookers (Matsushita and Sanyo), air conditioners (Mitsubishi), and
microwave ovens (Sharp, Sanyo, and Toshiba), to name a few. Matsushita used
fuzzy technology to develop its digital image stabilizer for camcorders. Adaptive
fuzzy systems (a hybrid with neural networks) are found in many Japanese cars.
Nissan has patented a fuzzy automatic transmission that is now very popular with
many other cars such as Mitsubishi and Honda [52].

3.4.1 Fuzzy Set Theory

The human thought, reasoning, and decision-making process is not crisp. We use
vague and imprecise words to explain our thoughts or communicate with one
another. There is a contradiction between the imprecise and vague process of
human reasoning, thinking, and decision-making and the crisp, scientific reasoning
of black and white computer algorithms and approaches. This contradiction has
given rise to an impractical approach of using computers to assist humans in the
decision-making process, which has been the main reason behind the lack of
success for conventional rule-based systems, also known as expert systems.[6] Expert
systems as a technology started in early 1950s and remained in the research lab-
oratories and never broke through to consumer market.

---

[6]For decades, the term "Artificial Intelligence" was synonymous with rule-based expert systems
due to some historical event dating back to early 1950s. This was the reason that until early 2000s
scientists, professionals, and practitioners hesitated to use the term AI (Artificial Intelligence) to
refer to their activities.

In essence, fuzzy logic provides the means to compute with words. Using fuzzy logic, experts no longer are forced to summarize their knowledge to a language that machines or computers can understand. What traditional expert systems failed to achieve finally became reality with the use of fuzzy expert systems. Fuzzy logic comprises of fuzzy sets, which are a way of representing nonstatistical uncertainty and approximate reasoning, which includes the operations used to make inferences [54].

Fuzzy set theory provides a means for representing uncertainty. Uncertainty is usually either due to the random nature of events or due to imprecision and ambiguity of information we have about the problem we are trying to solve. In a random process, the outcome of an event from among several possibilities is strictly the result of chance. When the uncertainty is a product of randomness of events, probability theory is the proper tool to use. Observations and measurements can be used to resolve statistical or random uncertainty. For example, once a coin is tossed, no more random or statistical uncertainty remains.

Most uncertainties, especially when dealing with complex systems, are the result of a lack of information. The kind of uncertainty that is the outcome of the complexity of a system is the type of uncertainty that rises from imprecision, from our inability to perform adequate measurements, from a lack of knowledge, or from vagueness (like the fuzziness inherent in natural language). Fuzzy set theory is a marvelous tool for modeling the kind of uncertainty associated with vagueness, with imprecision, and/or with a lack of information regarding a particular element of the problem at hand [55].

Fuzzy logic achieves this important task through fuzzy sets. In crisp sets, an object either belongs to a set or it does not. In fuzzy sets, everything is a matter of degrees. Therefore, an object belongs to a set to a certain degree. For example, the price of oil today is $52.69 per barrel.[7] Given the price of oil in the past few months, this price seems to be quite low.[8] But what is a low price for oil? A few months ago, the price of oil was about $100.00 per barrel.

The Paris-based International Energy Agency estimates that oil from shale formations costs $50–$100 a barrel to produce, compared with $10–$25 a barrel for conventional supplies from the Middle East and North Africa [56]. Taking into account the costs to produce a barrel of oil in the United States, one can say that today's price is low. If we arbitrarily decide that the cutoff for the "low" category of oil price is $55.00, and use crisp sets, then $55.01 is not a low oil price. However, imagine if this was the criterion that was used by oil company executives to make decisions. Imagine the number of layoffs that would be just around the corner. Fuzzy logic proposes the fuzzy sets for the price of oil shown in Fig. 3.15.

The most popular form of representing fuzzy set and membership information is shown in Eq. (3.1):

Mathematical representation of Fuzzy Membership Function.

---

[7]This is the price of oil on Saturday January 3, 2015.
[8]Price of oil plunged by about 50 % during the last couple of months of 2014.

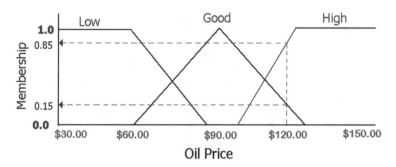

Fig. 3.15 Fuzzy sets representing the price of oil

$$\mu_A(x) = m \tag{3.1}$$

This representation provides the following information: the membership μ of x in fuzzy set A is m. According to the fuzzy sets shown in Fig. 3.15, when the price of oil is \$120.00 per barrel, it has a membership of 0.15 in the fuzzy set "Good" and a membership of 0.85 in the fuzzy set "High." Using the notation shown in Eq. (3.1) to represent the oil price membership values, notation in Eq. (3.2) can be used.

Mathematical representation of Fuzzy Membership Function for Oil Price.

$$\mu_{\text{Good}}(\$20.00) = 0.15 \quad \mu_{\text{High}}(\$20.00) = 0.85 \tag{3.2}$$

3.4.2 Approximate Reasoning

When decisions are made based on fuzzy linguistic variables ("Low," "Good," "High") using fuzzy set operators ("And," "Or"), the process is called "Approximate Reasoning." This process mimics the human expert's reasoning process much more realistically than the conventional expert systems. For example, if the objective is to build a fuzzy expert system to help us make a recommendation on enhanced recovery operations, then we can use the oil price and the company's proven reserves to make such a recommendation. Using the fuzzy sets in Fig. 3.15 for the oil price and the fuzzy sets in Fig. 3.16 for the company's total proven reserves, we try to build a fuzzy system that can help us in making a recommendation on engaging in enhanced recovery operations as shown in Fig. 3.17.

The approximate reasoning is implemented through fuzzy rules. A fuzzy rule for the system being explained here can have the following form:

Rule #1: IF the **Price of Oil** is **"High"** AND the **Total Proven Reserves** of the company is **"Low"** THEN **Engaging in Enhanced Recovery** practices is **"Highly Recommended."**

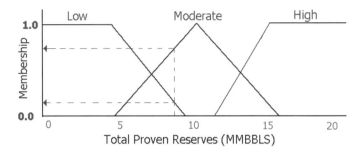

Fig. 3.16 Fuzzy sets representing the total proven reserves

Since this fuzzy system is comprised of two variables and each of the variables consists of three fuzzy sets, the system will include (3^2) nine fuzzy rules. These rules can be set up in a matrix as shown in Fig. 3.18.

The abbreviations that appear in the matrix above (Fig. 3.18) correspond to the fuzzy sets defined in Fig. 3.17. As one can conclude from the above example, the number of rules in a fuzzy system increases dramatically with addition of new variables. Adding one more variable consisting of three fuzzy sets to the above example, increases the number of rules from (3^2) nine to $(3^3) = 27$. This is known as the "curse of dimensionality."

3.4.3 Fuzzy Inference

A complete fuzzy system includes a fuzzy inference engine. The fuzzy inference helps build fuzzy relations based on the fuzzy rules that have been defined. During a fuzzy inference process, several fuzzy rules will be fired in parallel. The parallel rule firing, unlike the sequential evaluation of the rules in the conventional expert systems, is much closer to the human reasoning process. Unlike the sequential process where some information contained in the variables may be overlooked due to the stepwise approach, the parallel firing of the rules allows consideration of all the information content, simultaneously.

There are many different fuzzy inference methods. One of the popular methods is called the Mamdani's inference method [57]. This inference method is demonstrated graphically in Fig. 3.19. In this figure, a case is considered when the price of

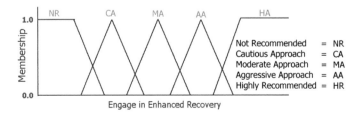

Fig. 3.17 Fuzzy sets representing the decision to engage in enhance recovery

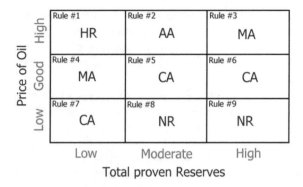

Fig. 3.18 Fuzzy rules for approximate reasoning

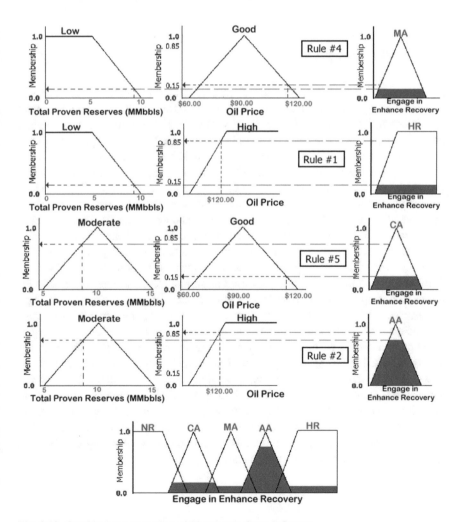

Fig. 3.19 Graphical representation of Mamdany's fuzzy inference

oil is $120.00 per barrel and the company has approximately 9 MMbbls of proven reserves. The oil price is represented by its membership in fuzzy sets "Good" and "High" while the total proven reserves is represented in fuzzy sets "Low" and "Moderate." As shown in Fig. 3.18, this causes four rules to be fired simultaneously.

According to Fig. 3.18 these are rules #1, #2, #4, and #5. In each rule, the fuzzy set operation "And" the intersection between the two input (antecedents) variables, is evaluated as the minimum and consequently is mapped on the corresponding output (consequent). The result of the inference is the collection of the different fuzzy sets of the output variable as shown on the bottom of the figure.

A crisp value may be extracted from the result as mapped on the output fuzzy sets by defuzzifying the output. One of the most popular defuzzification procedures is to find the center of the mass of the shaded area in the output fuzzy sets.

3.5 Evolutionary Optimization

Evolutionary Optimization, like other AI and data-driven analytics algorithms, has its roots in nature. It is an attempt to mimic the evolutionary process using computer algorithms and instructions. However, why would we want to mimic the evolution process? The answer will become obvious once we realize what type of problems the evolution process solves and whether we would like to solve similar problems. Evolution is an optimization process [58]. One of the major principles of evolution is heredity. Each generation inherits the evolutionary characteristics of the previous generation and passes those same characteristics to the next generation. These characteristics include those of progress, growth, and development. This passing of the characteristics from generation to generation is facilitated through genes.

Since the mid-1960s, a set of new analytical tools for intelligent optimization have surfaced that are inspired by the Darwinian evolution theory. The term "evolutionary optimization" has been used as an umbrella for many of these tools. Evolutionary optimization comprises of evolutionary programming, genetic algorithms, evolution strategies, and evolution computing, among others. These tools (and names) look similar and their names are associated with the same meaning to many people. However, these names carry quite distinct meanings to the scientists deeply involved in this area of research. Evolutionary programming, introduced by Koza [59], is mainly concerned with solving complex problems by evolving sophisticated computer programs from simple, task-specific computer programs. In evolution strategies [60], the components of a trial solution are viewed as behavioral traits of an individual, not as genes along a chromosome, as implemented in genetic algorithms. Evolution programs [61] combine genetic algorithms with specific data structures to achieve their goals.

3.5.1 Genetic Algorithms

Darwin's theory of survival of the fittest (presented in his 1859 paper titled: "On the Origin of Species by Means of Natural Selection"), coupled with the selectionism of Weismann and the genetics of Mendel, have formed the universally accepted set of arguments known as the evolution theory [60]. In nature, the evolutionary process occurs when the following four conditions are satisfied [59]:

- An entity has the ability to reproduce.
- There is a population of such self-reproducing entities.
- There is some variety among the self-reproducing entities.
- This variety is associated with some difference in ability to survive in the environment.

In nature, organisms evolve as they adapt to dynamic environments. The "fitness" of an organism is defined by the degree of its adaptation to its environment. The organism's fitness determines how long it will live and how much of a chance it has to pass on its genes to the next generation. In biological evolution, only the winners survive to continue the evolutionary process. It is assumed that if the organism lives by adapting to its environment, it must be doing something right. The characteristics of the organisms are coded in their genes, and they pass their genes to their offspring through the process of heredity. The fitter an individual, the higher is its chance to survive and hence reproduce.

Intelligence and evolution are intimately connected. Intelligence has been defined as the capability of a system to adapt its behavior to meet goals in a range of environments [60]. By imitating the evolution process using computer instructions and algorithms, researchers try to mimic the intelligence associated with the problem-solving capabilities of the evolution process. As in real life, this type of continuous adaptation creates very robust organisms. The whole process continues through many "generations," with the best genes being handed down to future generations. The result is typically a very good solution to the problem. In computer simulation of the evolution process, genetic operators achieve the passing on of the genes from generation to generation.

These operators (crossover, inversion, and mutation) are the primary tools for spawning a new generation of individuals from the fit individuals of the current population. By continually cycling these operators, we have a surprisingly powerful search engine. This inherently preserves the critical balance needed with an intelligent search: the balance between exploitation (taking advantage of information already obtained) and exploration (searching new areas). Although simplistic from a biologist's viewpoint, these algorithms are sufficiently complex to provide robust and powerful search mechanisms.

3.5.2 Mechanism of a Genetic Algorithm

The process of genetic optimization can be divided into the following steps:

1. Generation of the initial population.
2. Evaluation of the fitness of each individual in the population.
3. Ranking of individuals based on their fitness.
4. Selecting those individuals to produce the next generation based on their fitness.
5. Using genetic operations, such as crossover, inversion and mutation, to generate a new population.
6. Continue the process by going back to step 2 until the problem's objectives are satisfied.

The initial population is usually generated using a random process covering the entire problem space. This will ensure a wide variety in the gene pool. Each problem is encoded in the form of a chromosome. Each chromosome is collection of a set of genes. Each gene represents a parameter in the problem. In classical genetic algorithms, a string of "0"s and "1"s or a bit string represents each gene (parameter). Therefore, a chromosome is a long bit string that includes all the genes (parameters) for an individual. Figure 3.20 shows a typical chromosome as an individual in a population that has five genes. Obviously, this chromosome is for a problem that has been coded to find the optimum solution using five parameters.

The fitness of each individual is determined using a fitness function. The goal of optimization is usually to find a minimum or a maximum. Examples of this include the minimization of error for a problem that must converge to a target value or the maximization of hydrocarbon production in a shale well. Once the fitness of each individual in the population is evaluated, all the individuals will be ranked. After the ranking, it is time for selection of the parents that will produce the next generation of individuals. The selection process assigns a higher probability of reproduction to the highest-ranking individual, and the reproduction probability is reduced with a reduction in ranking.

After the selection process is complete, genetic operators such as crossover, inversion, and mutation are incorporated to generate a new population. The evolutionary process of survival of the fittest takes place in the selection and reproduction stage. The higher the ranking of an individual, the higher the chance for it to reproduce and pass on its gene to the next generation.

In crossover, the two parent individuals are first selected and then a break location in the chromosome is randomly identified. Both parents will break at that

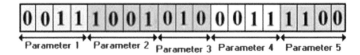

Fig. 3.20 A chromosome with five genes

Fig. 3.21 Simple crossover operator

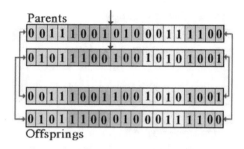

Fig. 3.22 Double crossover operator

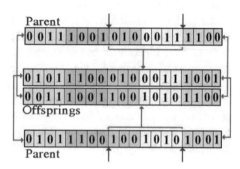

location and the halves switch places. This process produces two new individuals from the parents. One pair of parents may break in more than one location at different times to produce more than one pair of offspring. Figure 3.21 demonstrates the simple crossover.

There are other crossover schemes besides simple crossover, such as double crossover and random crossover. In double crossover, each parent breaks in two locations, and the sections are swapped. During a random crossover, parents may break in several locations. Figure 3.22 demonstrates a double crossover process.

As was mentioned earlier, there are two other genetic operators in addition to crossover. These are inversion and mutation. In both of these operators the offspring is reproduced from one parent rather than a pair of parents. The inversion operator changes all the "0"s to "1"s and all the "1"s to "0"s from the parent to make the offspring. The mutation operator chooses a random location in the bit string and changes that particular bit. The probability for inversion and mutation is usually lower than the probability for crossover. Figures 3.23 and 3.24 demonstrate inversion and mutation. In these figures each parameters is identified by a color.

Fig. 3.23 Inversion operator

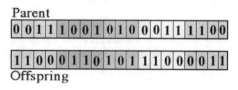

Fig. 3.24 Mutation operator

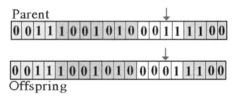

Once the new generation has been completed, the evaluation process using the fitness function is repeated and the steps in the aforementioned outline are followed. During each generation, the top ranking individual is saved as the optimum solution to the problem. Each time a new and better individual is evolved, it becomes the optimum solution. The convergence of the process can be evaluated using several criteria. If the objective is to minimize an error, then the convergence criteria can be the amount of error that the problem can tolerate. As another criterion, convergence can take place when a new and better individual is not evolved within 4–5 generations. Total fitness of each generation has also been used as a convergence criterion. Total fitness of each generation can be calculated (as a sum) and the operation can stop if that value does not improve in several generations. Many applications simply use a certain number of generations as the convergence criterion.

As you may have noticed, the above procedure is called the classic genetic algorithms. Many variations of this algorithm exist. For example, there are classes of problems that would respond better to genetic optimization if a data structure other than bit strings were used. Once the data structure that best fits the problem is identified, it is important to modify the genetic operators such that they accommodate the data structure. The genetic operators serve specific purposes, making sure that the offspring is a combination of parents in order to satisfy the principles of heredity, which should not be undermined when the data structure is altered.

Another important issue is introduction of constraints to the algorithm. In most cases, certain constraints must be encoded in the process so that the generated individuals are "legal." Legality of an individual is defined as its compliance with the problem constraints. For example in a genetic algorithm that was developed for the design of new cars, basic criteria, including the fact that all four tires must be on the ground, had to be met in order for the design to be considered "legal." Although this seems to be quite trivial, it is the kind of knowledge that needs to be coded into the algorithm as constraints in order for the process to function as expected.

3.6   Cluster Analysis

Cluster analysis, by nature is an unsupervised process. It aims at discovering order and patterns in seemingly chaotic, hyper-dimensional data. Cluster analysis is the unsupervised classification of data into groups. The classification is made based on the similarities of the data records to one another, such that similar data records will be classified in the same group. One of the objectives of data-driven analytics is making the maximum use of the information content of a datasets. Information content is directly related to the order in data. In other words, if one can identify order in data, the "order" will guide the analyst toward the information content. Cluster analysis is one way to discover order in data. Most cluster analysis techniques are unsupervised.

Most clustering algorithms do not rely on assumptions common to conventional statistical methods, such as the underlying statistical distribution of data, and therefore they are useful in situations where little prior knowledge exists. The potential of clustering algorithms to reveal the underlying structures in data can be exploited in a wide variety of applications, including classification, image processing, pattern recognition, modeling, and identification [62].

The opposite of "order" is "chaos." Large datasets especially when they include a large number of variables (high dimensions) tend to look chaotic. Humans cannot plot or imagine in more than three dimensions. Sometimes innovative techniques can be used in order to be able to visualize data in four and sometimes even five dimensions. This can be achieved by plotting the data in three dimensions and then showing the magnitude of another dimension with the size or the color of the points being plotted, etc. Nevertheless, as the number of records and parameters (rows and columns in a flat file) in a dataset increases, visualization, with the objective of discovering trends and patterns (order) in data, becomes more and more difficult and less productive.

Clustering algorithms are developed to help with discovering "order" in data, especially with large datasets, where the effect of visualization on discovering patterns and trends start diminishing. There are a couple of items regarding the algorithms that perform cluster analyses that needs to be paid attention to. First is the selection of the dimensions that are to be used and the second is the number of clusters. When a clustering algorithm is to be applied to a dataset, these two items need to be provided. This is a set of information that we usually do not have in advance and need to use domain expertise and/or "trial and error" in order to get it right. The other technique would be to try a very large number of combinations of these two variables (number and combination of the dimensions [parameters] involved and the number of clusters) and then identify the best clustering result. Obviously, for each dataset there is an optimum number and combination of parameters and an optimum number of clusters that can release (percolate) maxim amount of information from a dataset. However, finding these optimum numbers of clusters and combination of parameters is the key.

The next question is "how can we decide if one result of cluster analysis is better than the other?" The answer is: "the cluster analysis that generates better order (less chaos) is the better analysis." But how can we measure order? Well, we can measure chaos using entropy (the higher the degree of the chaos, the higher is the entropy). Therefore, the analysis that results in less entropy should have higher order and therefore should be the better analysis. Entropy is defined as a measure of

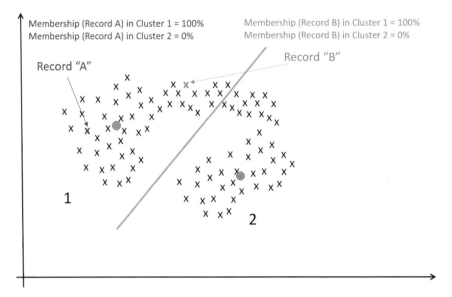

Fig. 3.25 In cluster analysis, clusters are separated via crisp lines and two-valued logic is used for clustering

the disorder in a system. Entropy is a state function. Shannon's entropy [47] is usually used to measure entropy of a system.

Upon completion of Cluster Analysis each record in the dataset belongs to one of the multiple clusters that have been identified (Fig. 3.25). The existence of the data record in a given cluster determines its similarity to the cluster center. Therefore, all data records in a cluster share certain characteristics that separate them from the data records that belong to other clusters.

3.7   Fuzzy Cluster Analysis

In conventional cluster analysis as shown in Fig. 3.25 clusters are separated by crisp boundaries. In this figure the two data points that are identified by the red cross and the green cross belong to cluster "1" and do not belong to cluster "2." In this figure cluster centers are identified by brown circles. In Fig. 3.25, both identified data points have a membership of "1" in cluster "1" and a membership of "0" in cluster "2."

If Fig. 3.25 was not observable (for example instead of two, it was part of a hyper-dimensional dataset) and you would only be exposed to the algorithm output, then you would assume that these two points are quite similar. For example, if the cluster centers were representative of rock qualities (1 = Good Shale and 2 = Poor Shale), both these wells were completed in "Good" quality shale. However, the reality, as presented in Fig. 3.25 is quite different from this interpretation.

Fuzzy Cluster Analysis [63] that is an implementation of Fuzzy Set Theory [51] in cluster analysis was introduced several years ago. In accordance with the spirit of Fuzzy Set Theory, Fuzzy Cluster Analysis indicates that membership in a cluster is not defined as 0 or 100 %. Membership of a record in a cluster is based on fuzzy membership function. Each record is part of all the clusters, but it belongs to each cluster to a degree. This is an extension of the Cluster Analysis that allows data record to be partially member of multiple clusters, rather than belonging to one cluster 100 % and to other clusters 0 %.

This is clearly demonstrated in Fig. 3.25 where two clusters "1" and "2" are separated by a crisp, hard line. As mentioned previously, two records identified as "Record A" and "Record B" are both in cluster 1. These two records share the same membership in both clusters 1 and 2. Both records belong to Cluster 1, 100 % and to Cluster 2, 0 %. In other words their membership in cluster 1 is 1 and in cluster 2 is 0. A closer look at the "Record A" and "Record B" in Fig. 3.25 shows that although these two records are located in Cluster "1," they are different. In other words, by clustering "Record A" and "Record B" into one cluster we have indeed release "some" information content from this dataset, but is there more information that we can extract from this dataset regarding these two records? The answer is "yes."

Figure 3.26 shows the same dataset as Fig. 3.25, however, in Fig. 3.26 clusters "1" and "2" are not separated by a crisp boundary. While keeping the cluster centers

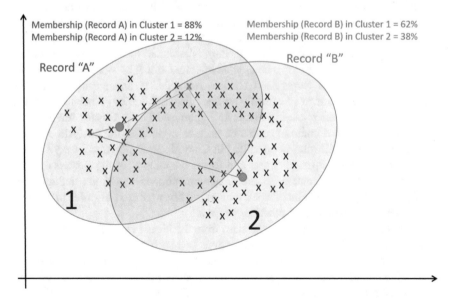

Fig. 3.26 In fuzzy cluster analysis, clusters are no longer separated via crisp lines and are identified with multivalued logic (Fuzzy Logic) in order to discover order in the data

in Cluster "1" and Cluster "2" the same, in the algorithm that is presented in this figure, we assign a partial membership to each record for each cluster. When comparing Record "A" with Record "B," we can see that Record "A" is closer to Cluster center "1" than the Cluster center "2." The same is true about Record "B" (that is why they were both allocated to Cluster "1" in Fig. 3.25), however, the degree of closeness to the cluster centers for these records is not the same. They both belong to Cluster "1" more than they belong to Cluster "2" but this belonging is not the same as the crisp cluster analysis suggests.

When using Fuzzy Cluster Analysis we can assign partial membership to each record for each cluster. For example, according to Fig. 3.26, Record "A" belongs to Cluster "1" with a membership of 0.88 and belongs to Cluster "2" with a membership of 0.12. Furthermore, Record "B" belongs to Cluster "1" with a membership of 0.62 and belongs to Cluster "2" with a membership of 0.38. This shows that Fuzzy Cluster Analysis releases more information about the dataset when it is compared to conventional Cluster Analysis, and in the view of the author it is better than conventional Cluster Analysis.

3.8   Supervised Fuzzy Cluster Analysis

As was mentioned before, Cluster Analysis (whether it be crisp or fuzzy) is an unsupervised data mining technology. Its aim is to analyze the dataset and attempt to release as much information content from the dataset as it can. In the cluster analysis algorithms the objective is to find the optimum location of the cluster centers for as many clusters as the algorithm is assigned to do. Supervised Fuzzy Cluster Analysis (this could also be applied to conventional cluster analysis and the result would be called Supervised Cluster Analysis) is an algorithm that was developed by the author to specifically address the peculiarities associated with the analysis of the datasets related to Shale. In Supervised Fuzzy Cluster Analysis the objective is to allow domain expertise be used as a guide in the analysis. A good example of such application is presented in Chap. 5. In Supervised Fuzzy Cluster Analysis the domain expert identifies and forces the number of the clusters and the location of the cluster centers. In the chapter mentioned above the definition of "Poor," "Average," and "Good" rock (shale) quality was used in order to analyze the impact of the completion practices on production.

Several new types of analyses can stem from the ideas behind Supervised Fuzzy Cluster Analysis. In the following two sections two such analyses are presented. Of course in order to use Supervised Fuzzy Cluster Analysis, one needs to have access to a software application that accommodates such interaction between domain experts and the data mining algorithms. Figures 5.2 and 5.3 are examples of using Supervised Fuzzy Cluster Analysis in shale related data mining analysis.

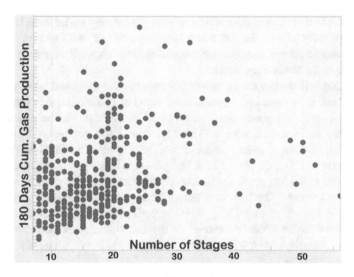

Fig. 3.27 This graph represent the cross plot of 180 days cumulative gas production versus total number of stages for more than 130 Marcellus well

3.8.1 Well Quality Analysis (WQA)

As a new application of Supervised Fuzzy Cluster Analysis in the upstream oil and gas industry, Well Quality Analysis (WQA) is a new and innovative algorithm that has been developed specifically for the data-driven analysis of oil and gas related problems.[9] In Well Quality Analysis we classify production from each well using fuzzy sets. In other words, we allow the boundary of the definition of different well qualities to overlap. Let us use an example to clarify this method. Figure 3.27 shows the "180 days cumulative production" of more than 130 wells completed in Marcellus shale. This figure shows the total number of stages that was used to complete each of the wells with the corresponding production indicator (180 days cumulative production) for each well. It is important to note that the degree of scatter in this plot. It is hard to make out any trends or patterns from this data. The objective of Well Quality Analysis is to see if any patterns or trends can be discovered and displayed from this seemingly chaotic behavior.

To perform these analyses, we first divide the wells into classes. In the first attempt we divide the wells into three classes[10] of "Poor," "Average," and "Good" wells. However, unlike the artificial, crisp classification we use a natural, fuzzy definition for each well quality. We call the crisp classification as artificial, because it is different with the way human brain (the standard for natural way of doing

---

[9]Use of these analysis in problems related to other industries, of course, is quite possible. In this analysis since we are working with wells, we chose to use the name "Well Quality Analysis".

[10]Later, it will be shown that then number of classes will be increased to four and then five.

thing) works in classifying objects. For example, the reality is that there is not a Mcf of gas that really makes the difference between a "Poor" well and an "Average" well. No matter where you put the separation line, such classification is simply unrealistic and actually ridiculous.

For example if we use the boundary between "Poor" wells and "Average" wells to be the "180 days cumulative production" value of 100 MMcf, then a well with a "180 days cumulative production" value of 99.999 MMcf will be classified as a "Poor" well while a well with a "180 days cumulative production" of 100.001 MMcf will be classified as an "Average" well. In this classification less than 2 Mcf of production in 180 days (about 11 ft^3 of gas—this is below the precision of some measurement tools) has made the difference between an entire class of wells. This example demonstrates the ludicrous nature of crisp (or conventional) cluster analysis when it applies to classifying wells to perform meaningful analysis. By the way, it is notable that almost everyone[11] in the data analytics arena (in the upstream oil and gas industry) uses conventional cluster analysis when classifying hydrocarbon producing wells.

The top graph in Fig. 3.28 shows that the wells in this Marcellus shale asset are divided into three fuzzy sets identified as "Poor," "Average," and "Good" wells. The middle graph in Fig. 3.28 shows that the same set of well are divided into four fuzzy sets identified as "Poor," "Average," "Good," and "Very Good" wells. The bottom graph in Fig. 3.28 shows that the same set of well are divided into five fuzzy sets identified as "Poor," "Average," "Good," "Very Good," and "Excellent" wells. The idea that needs to introduce here is the idea of granularity. Granularity is defined as the scale and/or the level of detail that is presented in an analysis or in a set of data. We will visit this idea again in the next section. Figure 3.28 shows that the granularity in defining and classifying well qualities in this analysis has changed from three to five classes of wells.

To perform the Well Quality Analysis, we first classify all the wells in this Marcellus shale asset based on the definitions that are demonstrated in Fig. 3.28. Then we use the membership functions of each well in order to calculate the contribution of the parameter that we are trying to analyze (in this example, the total number of stages). For example, the total number of stages from a well that is classified as being 100 % poor, will be allocated to the class of all "Poor" wells, while the total number of the stages of a well that has been identified as partially "Poor" and partially "Average" will be allocated to each category based on this membership function. For example the well shown in the top graph of Fig. 3.28 has a membership of 0.77 in the class of "Poor" wells and membership of 0.23 in the class of "Average" wells. Therefore, total number of stages for this well will be allocated accordingly.

Applying the above algorithm to all the wells in this asset (with three classes) will result in the bar chart shown in Fig. 3.29. This figure shows that the average

---

[11]That is everyone other than Intelligent Solutions, Inc. that has developed this algorithm for Well Quality Analysis.

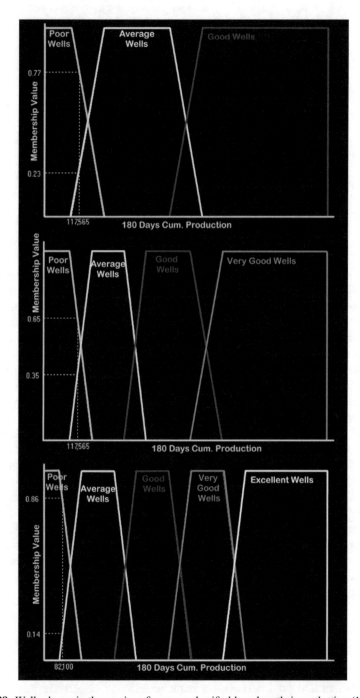

Fig. 3.28 Wells shown in the previous figure are classified based on their production (180 days cum. production) value. Same number of wells can be divided into three (*top*), four (*middle*), or five different classes (*bottom*)

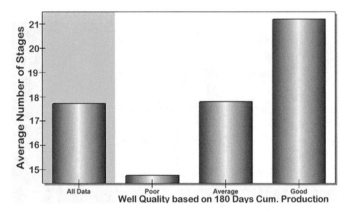

Fig. 3.29 Well Quality Analysis applied to more than 130 wells in the Marcellus shale shows that the wells with higher quality have been completed with more number of stages

number of stages for all wells in this asset is about 18 stages per wells (the bar in the figure with pink background). However, the trend and pattern is unmistakable (with the three bars in yellow background) that "Poor" wells have been completed with an average of less than 15 stages, while "Average" wells and "Good" wells have been completed with an average of 18 stages and 21 stages, respectively. There is nothing unclear about the pattern that is displayed in Fig. 3.29.

As the granularity of the analysis is increased from three to four and then to five classes (as shown in middle and bottom graphs in Fig. 3.28), the general behavior of the result inferred in Fig. 3.29 will hold. This is clearly shown in Fig. 3.30. In this figure it is shown that when four classes are used to analyze the contribution of the total number of stages to productivity of wells in this asset, then the "Poor" wells have been completed with an average of less than 14 stages, while "Average" wells, "Good" wells, and "Very Good" wells have been completed with an average of 18 stages, 20 stages, and 22 stages, respectively. Figure 3.30 includes the results of the analysis when the granularity of the analysis is increased to five classes.

Figures 3.27, 3.28, 3.29, and 3.30 clearly demonstrate the power of fuzzy set theory when it is incorporated into the Well Quality Analysis in discovering and presenting hidden trends and patterns that are embedded in the raw data and that can be extracted and displayed. In the next section we take the idea presented in this section one step further in order to show how continuous trends can be extracted from seemingly chaotic and scatter data.

3.8.2 Fuzzy Pattern Recognition

Fuzzy Pattern Recognition is another implementation of Supervised Fuzzy Cluster Analysis that is combined with an optimization algorithm in order to discover

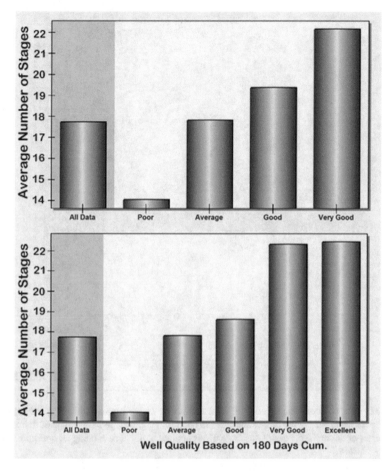

Fig. 3.30 Well Quality Analysis applied to the wells in the Marcellus shale using four and five fuzzy sets

trends in seemingly chaotic data. Fuzzy Pattern Recognition as it is implemented and explained here can be thought of as an extension of the Well Quality Analysis that was explained in the previous section. If we increase the granularity of the analysis that was explained in the previous section to its maximum (an optimum value that shows a trend, clearly), then instead of three, four, or five classes, we can have a large number of classes that can provide a continuous trend that results in a line rather than a bar chart. Figure 3.31 shows the application of Fuzzy Pattern Recognition to total number of stages and the production indicator (180 days of cumulative production).

Two items need to be emphasized regarding this figure. The gray dots that have also been plotted in the same graph shows the actual data that was used to generate

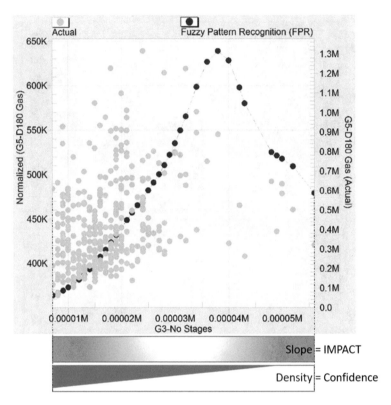

Fig. 3.31 Fuzzy Pattern Recognition of total number of stages versus production indicator. The trend shown using the *purple points* and *line* are not a moving average or any type of regression. It is the result of Supervised Fuzzy Cluster Analysis applied to the dataset using an optimum number of fuzzy clusters

the Fuzzy Pattern Recognition curve. The slope of the purple line determines the impact of the *x* axis (total number of stages) on the *y* axis (180 days of cumulative production) while the density of the (gray) data points on this plot indicates the degree of confidence that can be expressed about the trend being shown by the purple Fuzzy Pattern Recognition curve. The two strips of colors on the bottom of Fig. 3.31 are presented to show the degree of confidence and the impact of the parameter.

Figures 3.32, 3.33, 3.34, 3.35, 3.36, 3.37, 3.38, and 3.39 are presented in order to graphically explain the steps involved in the analysis of the Fuzzy Pattern Recognition curve. Figure 3.32 shows the histogram distribution of the independent variable (total number of stages) that is being analyzed. This figure also shows the overlay of the histogram distribution of the independent variable on the Fuzzy

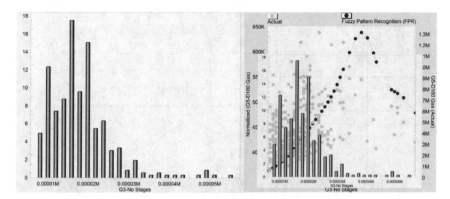

Fig. 3.32 Histogram distribution of total number of stages in the dataset (on the *left*) along with its overlay on the Fuzzy Pattern Recognition (on the *right*) to clarify the level of confidence on the impact analysis

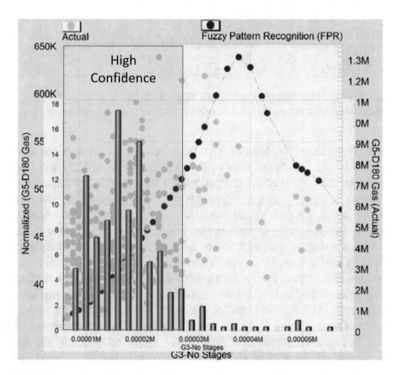

Fig. 3.33 Overlay of the histogram distribution of the total number of stages on Fuzzy Pattern Recognition, identifying the portion of the impact that carries high confidence

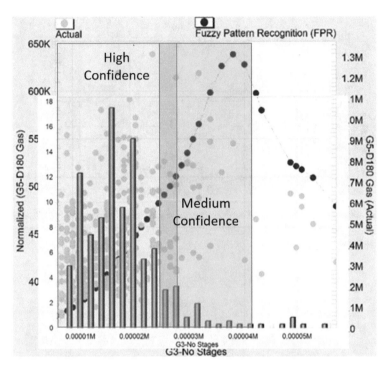

Fig. 3.34 Overlay of the histogram distribution of the total number of stages on Fuzzy Pattern Recognition, identifying the portion of the impact that carries high and medium confidence

Pattern Recognition curve. Figures 3.33, 3.34, and 3.35 show how the histogram distribution of the density of the available data can help identify the degree of confidence on the impact of the independent variable on the dependent variable.

As shown in these figures, in the ranges of the independent variable that large number of data is present, higher confidence is expressed on the impact of the independent variable on the dependent variable. To stay loyal to the fuzzy nature of the analysis the degree of confidence in these figures are identified by semantics such as high, medium, and low and they overlap.

The slope of the Fuzzy Pattern Recognition curve is an indicator of the impact of the independent variable (the x axis—total number of stages) on the dependent variable (y axis—180 days of cumulative production). Figure 3.36 is a schematic of the impact of total number of stages on the 180 days of cumulative production as indicated in Fig. 3.31. Figure 3.37 divides the regions of impact in the figure to portions where the impact is high (large slope), low (small to no slope), and medium (small slope). In Fig. 3.38 the results of Figs. 3.32, 3.33, 3.34, and 3.35

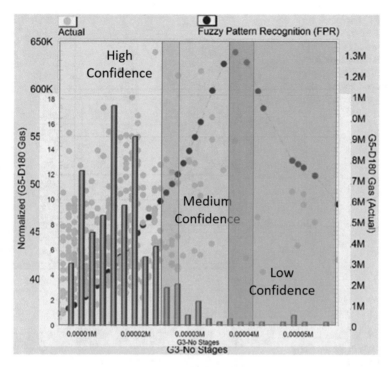

Fig. 3.35 Overlay of the histogram distribution of the total number of stages on Fuzzy Pattern Recognition, identifying the portion of the impact that carries high, medium, and low confidence

Fig. 3.36 Explanation of the Fuzzy Pattern Recognition Plot. Slope of the FPR curve shows the impact of the independent variable (*x* axis) on the dependent variable (*y* axis)

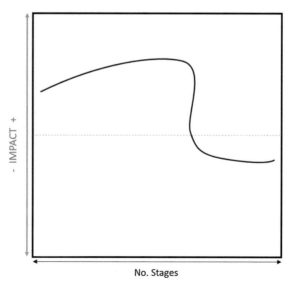

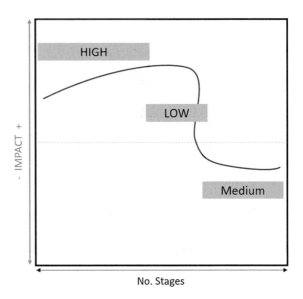

Fig. 3.37 Explanation of the Fuzzy Pattern Recognition Plot. Portions of the FPR curve that indicates high, medium, and low impact of the independent variable (*x* axis) on the dependent variable (*y* axis)

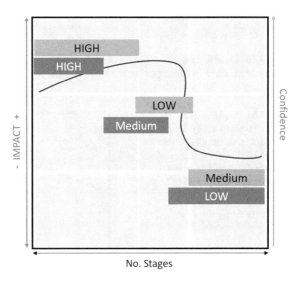

Fig. 3.38 Explanation of the Fuzzy Pattern Recognition Plot. Overlaying the confidence levels on the impact curve

that are the histogram distribution of the independent variable are superimposed on Fig. 3.37 to show the level of confidence on each section of the impact analysis.

Finally, Fig. 3.39 shows the outcome of the complete analysis. In this figure, two sets of qualifying semantics regarding the impact of the independent variable (total number of stages) on the dependent variable (180 days of cumulative production) are displayed. As shown in this figure, the graph is divided into portions that indicate high impact along with high confidence (left hand side of the plot) all the

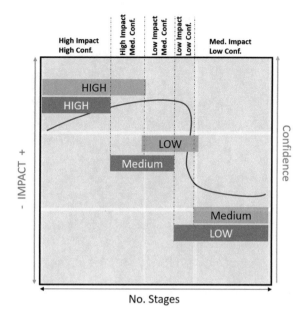

Fig. 3.39 Explanation of the Fuzzy Pattern Recognition Plot. Results of the combining the impact and confidence curves to analyze the Fuzzy Pattern Recognition Plot

way to portions of the plot that indicates low impact along with low confidence. The conclusion is that much can be learned from the Fuzzy Pattern Recognition curve, especially when it is compared with the raw data as is shown in Fig. 3.27.

Chapter 4
Practical Considerations

With data analytics becoming popular in the upstream oil and gas industry and finding its way into addressing some of the most important issues in the reservoir and production management of the unconventional resources such as shale, many big data analytics and data-driven modeling enthusiasts have decided to venture into the upstream oil and gas industry and test their talent in solving oil and gas-related problems. Some of these enthusiasts that have limited, if any, background in geosciences or petroleum engineering make the common mistake of treating the issues concerning our industry with those faced by social media, retail, or even pharmaceutical industry, and thus try purely statistical (or even machine learning) approaches that have worked for those industries, in the upstream oil and gas.

What is being completely overlooked in this process is the nature of the problems that are faced in our industry and more than seven decades of successful practices and lessons learned. Therefore, the attempt to reinvent the wheel (albeit using a different set of tools and approaches) by these non-domain experts, usually results in, at best, mediocre outcomes and sometimes even embarrassing ones. The processes that are of interest in our industry, while being hugely sophisticated and complex, have solid foundation in physics and geosciences. Therefore, lack of understanding of the physical and the geoscientific principles that control all issues in our industry, or at times, ignoring such principles, seriously handicaps any data-driven or data analytic efforts. In this chapter of the book, important issues that need to be considered during the application of big data analytics and data-driven modeling in oil and gas-related problems in general, and specifically when it comes to reservoir and production management in shale will be addressed.

© Springer International Publishing AG 2017
S.D. Mohaghegh, *Shale Analytics*, DOI 10.1007/978-3-319-48753-3_4

4.1 Role of Physics and Geology

Although the technologies presented in this book do not start from the first principles physics, they are physics-based (geoscience-based) models. However, the incorporation of the physics and geosciences in these technologies are quite non-traditional. Reservoir characteristics and geological aspects are incorporated in the models for as much as they can be measured; while interpretations (biases) are intentionally left out during the model development. Although fluid flow through porous media is not explicitly (mathematically) formulated during the development of data-driven models, successful development of such models, is unlikely without a solid understanding and experience in reservoir and production engineering as well as geosciences.

Physics and geology are the foundation and the framework for the assimilation of the datasets that are used to develop (train, calibrate, and validate) the data-driven models. Therefore, Shale Analytics is not a blind search of parameters that may or may not impact the reservoir and production behavior during which we must solely rely on statistical relationship, rather, it is a well-directed and guided search through all field measurements in order to capture the inter-relations between them with the understanding that sometimes we may not have access to the desired parameters (due to the difficulty of its measurement) and may have to identify (sometimes through a trial and error process) the best measured parameter or collection of parameters that can act as a proxy for the desired characteristics.

4.2 Correlation is not the Same as Causation

One of the most contentious issues that are usually brought up by engineers and scientist, when presented by statistics, is the relationship between correlation and causation and the fact that they are not necessarily the same. In other words, just because two variables correlate, does not mean that one is the cause of the other. Figures 4.1, 4.2 and 4.3 are three examples that clearly demonstrate these phenomena.

Data presented in these figures are from public databases. Figure 4.1 shows how U.S. spending on science, space, and technology (in millions of dollars) correlates with the number of suicides by hanging, strangulation, and suffocations. The correlation is a staggering 0.992. However, there should be little doubt that these two phenomena have absolutely nothing to do with one another. This is a clear example where correlation has nothing to do with causation.

Similarly, in Fig. 4.2 divorce rate in the state of Maine is plotted as a function of time as well as per capita consumption of margarine. These two graphs show a correlation of 0.993 while having absolutely nothing to do with one another. In Fig. 4.3, it is demonstrated that U.S. crude import from Norway has a correlation

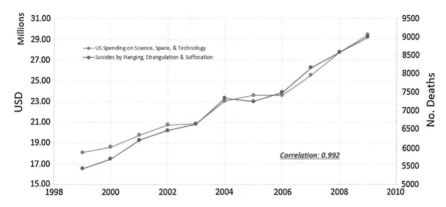

Fig. 4.1 A good example of lack of relationship between correlation and causation: U.S. spending on science, space and technology highly correlate with number of suicides by hanging, strangulation, and suffocation

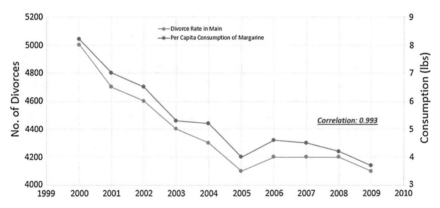

Fig. 4.2 Another good example of lack of relationship between correlation and causation: divorce rate in State of Maine highly correlates to per capita consumption of Margarine in the United States

coefficient of 0.955 with drivers killed in collisions with railway trains; another vivid example of lack of relationship between correlation and causation.

The objective of presenting the above examples is to emphasize the fact that when it comes to data-driven modeling, it is not enough to show correlation. The model must make physical and geological sense. Of course, importance of physics and geology only matters when a good correlation is achieved otherwise, the entire case is mute. In other words, once we have a model that can make good predictions, we can analyze its internal interactions between parameters in order to see if it actually makes physical and geological sense.

This is how data-driven modeling is distinguished with its competitor empirical technologies that rely purely on statistics and mathematics without regards for physics and geology. Furthermore, it addresses another important criticism that

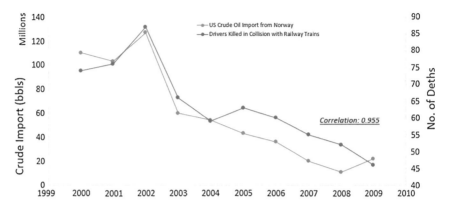

Fig. 4.3 Another good example of lack of relationship between correlation and causation: U.S. crude oil import from Norway highly correlates to number of drivers killed in collisions with railway trains

sometimes is bestowed upon data-driven technologies by those with superficial understanding of the technology. Data-driven modeling has been referred to as a "black box" technology. This comes from lack of understanding of the technology. If the term "black box" is used to emphasize the fact that there is not a deterministic mathematical formulation that can fully and comprehensively explain the behavior of a data-driven model, then this term is correct. However, if it means that the functionality of the model cannot be understood or verified, then it is a misuse of the term. All interactions between different parameters in a trained data-driven model can be fully investigated in order to make sure that it makes physical and geological sense. From this point of view, there is nothing in the data-driven model that makes it a "black box."

4.3 Quality Control and Quality Assurance of the Data

Let us face it. Real data is noisy. The only time you can expect clean data is when computer simulation models are used to generate the data. Whenever, real data, field measurements are used in any situation, one must expect to have noise in the data. This fact must not deter us from using real data as the main source of information in order to build models. In the context of data-driven models, there are specific issues that need to be taken into account when data quality is being considered. For this purpose, a serious data quality control and quality assurance need to be implemented as an important part of the data-driven modeling.

Given the fact that Shale Analytics includes data-driven modeling technology, it should be obvious that Quality Control and Quality Assurance (QC/QA) of the data is of immense importance. The famous phrase that states "Garbage-In,

Garbage-Out" takes on new meaning during the data-driven modeling process since the foundation of the model is all about what we put in it (data).

However, there are misconceptions about Quality Control and Quality Assurance (QC/QA) when it comes to data-driven analytics specifically regarding its applications in reservoir and production modeling. In this context, QC/QA takes an engineering meaning above and beyond the normal quality control of the data. In data-driven modeling, the modeler is engaged in teaching a computer (machine) reservoir and/or production engineering with data as the only available tool. It is important to communicate reservoir and production engineering facts to the computer. A good example of this practice is communication of production behavior to the computer, since it usually is one of the most important outputs of a data-driven model used in Shale Analytics.

Let us use two examples in order to explain data Quality Control and Quality Assurance (QC/QA) when it comes to data-driven models. The first example will have to do with formation thickness and the second example is related to production values. In this context, we discuss the difference between "no data" (blank cell in a spreadsheet), and the value of "0" (zero). Unfortunately, some make the grave mistake of thinking these two are equivalent.

For example, when the parameter is formation thickness, the value of zero has a very specific meaning. Let us assume that we are developing a data-driven model for a field that is producing from two reservoirs R1 and R2. For each record in the database that refers to a month of production for a well in the filed, we need to include two values for the formation thickness of each of the reservoirs, one for R1 and one for R2 in "ft." If for one of the formation thicknesses, (let us say for R1) we put the value of "0," we are simply stating a fact that the reservoir R1 is not present (has pinched out) at the location where this particular well has been completed. Therefore, for this particular well, all the productions should be attributed to reservoir R2.

However, if instead of the value "0" the value for the reservoir R1 is left blank, it means that (for whatever reason) we do not know the formation thickness for this reservoir at the location of this well. Therefore, if there is no reason for us to believe that this reservoir is absent in this location, then we must identify a value for the formation thickness at this location. The value of "0" is as likely to be true in this location as any other value (given the general formation thickness in this filed). Therefore, the best solution in such a case would be to use geostatistics and populated the value of the formation thickness for this reservoir at this location.

Next we visit the data quality when it comes to production. Since we are dealing with real world and not academic problems, the production profile that is presented to the computer for training purposes to be used as the target for correlation is quite noisy. It is important to communicate the source of the observed noise with the computer in order to train the model on the types of noise that can be expected in production profile. In other words, we need to teach the computer that if no disturbances are involved during a production, a well will produce in a clean and well-behaved manner. Good examples of such behavior are Decline Curves, or production profile from a well generated by a numerical simulator. But one will

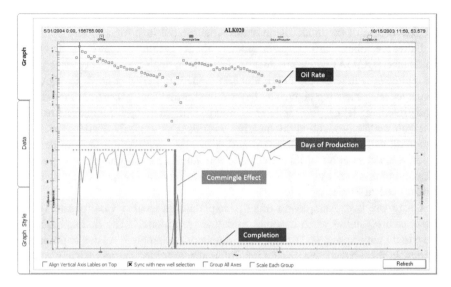

Fig. 4.4 Production behavior of a well completed in Persian Gulf. The noise in production (barrels per month—the *top graph*) can be traced by the inconsistencies in operational constraints such as, well head pressure (choke size), number of days of production as well as completion modifications

never see such behavior from an actual well. The top graph in Fig. 4.4 is a good example.

This figure includes actual oil production (in total barrels per month) profile from an offshore well in the Persian Gulf (top graph). There are two types of noises that can be observed in this graph. First one is mainly white noise that usually is the result of measurement inconsistencies. The result of such noise is a bit of up and down that does not disrupt the general trend in the production profile. However, a second type of noise, that is the subject of this discussion, is the noise that causes a change in the trend of production. For example in this figure, the second month of production is higher than the first month of production before a decline trend is started. Furthermore, halfway in the life of the well, the production is disrupted to almost zero before starting an increasing trend that takes production to a level even higher than when it was interrupted, before a new trend gets under way, and so forth.

It needs to be communicated during the training that this behavior is caused by human intervention (operational constraints) and is not the normal behavior of a well. The bottom graph in this figure includes two of the several possible graphs that can help explain the production profile behavior. In this graph, number of days of production per month and completion footage is plotted as a function of time (same *x* axis of production). This graph shows that during the first month, the well was only producing for a few days (well started production in the middle of the month), while starting the second month, the number of days started being more consistent. This explains the increase in the total oil production increase from the first to second month. Furthermore, it shows that in the middle of the profile, the well was shut in and then it was opened again. Furthermore, once it was put to production again, new parts of the formation were completed and are now contributing to production. This information contributes to understanding the behavior displayed in the oil production (top graph).

By providing these types of information during the training of the data-driven model, we are assisting the de-convolution of the information. The oil production that is presented in the top graph, that is typical of all real-life cases, is convoluted by the impact of multiple classes of factors such as well trajectory characteristics, formation (reservoir) characteristics, completion characteristics, and finally production characteristics.

By providing enough examples of a combination of all the above characteristics, we are helping the model to be able to distinguish between contribution of each of the involved parameters and eventually, the capability to de-convolve the information. This is the only way that the model will be able to perform reasonably in the future when it tries to perform production prediction.

Chapter 5
Which Parameters Control Production from Shale

Is it the quality of the formation or the quality of the completion that determines or controls the productivity of a shale well? Before the "Shale Revolution" one would never hear such a question. It was always a well-known fact that productivity of a well is controlled by reservoir characteristics. If you drill in a part of the reservoir with poor reservoir quality, then production will be poor. The entire idea of "sweet spot" has to do with identification of parts of the field with good reservoir characteristics. But then came production from shale. Since the completion practice in shale is unprecedented, some started asking the question "Is it the quality of the Formation or the Completion that controls production of a shale well"?

Let us use Shale Analytics to address this important question. This is a fit-for-purpose approach with no attempt to generalize the final conclusions. In other words, similar analyses must be performed for every shale asset before the question for a given asset (in a given shale play) can be answered. The analysis presented here is based on "Hard Data" or field measurements. No assumptions are made regarding the physics of the storage and/or the transport phenomena in shale. In this analysis, like all other applications of Shale Analytics, we let the data speak for itself.

The case study includes a large number of wells in a shale asset in the northeast of the United States. Characteristics such as net thickness, porosity, water saturation, and TOC are used to qualitatively classify the part of the formation that each well is producing from. Furthermore, wells are classified based on their productivity. We examine the "common sense" (conventional wisdom) hypothesis that reservoir quality has a positive correlation with well productivity (wells completed in shale with better reservoir quality will demonstrate better productivity). The data from the field will either confirm or dispute this hypothesis. If confirmed, then it may be concluded that completion practices have not harmed the productivity and are, in general, in harmony with the reservoir characteristics. The next step in the analysis is to determine the dominant trends in completion and judge them as best practices.

© Springer International Publishing AG 2017
S.D. Mohaghegh, *Shale Analytics*, DOI 10.1007/978-3-319-48753-3_5

However, if and when the hypothesis is disproved (wells completed in shale with better reservoir quality will *NOT* demonstrate better productivity), one can and should conclude that completion practices are the main culprit for the lack of better production from better quality shale. Since better reservoir quality provides the necessary precursor of better production, if it is not realized, then it must have to do with what was done to the well (during completion). In this case, analysis of the dominant trends in the completion practices should be regarded as identifying the practices that need to be modified (or be avoided, since what was done caused harm).

5.1   Conventional Wisdom

In this chapter of the book, using Shale Analytics, it is demonstrated that production from shale challenges many of our preconceived notions (conventional wisdom). Analyses presented in this chapter show that the impact of completion practices in low quality shale are quite different from those of higher quality shale. In other words, completion practices that results in good production in low quality shale are not necessarily just as good for higher quality shale. Results of this study will clearly demonstrate that when it comes to completion practices in shale, "One-Size-fit-All" is a poor prescription.

The conventional wisdom developed over several decades in the oil and gas industry states that better quality rocks produce more hydrocarbons. In other words, there is a positive correlation between reservoir characteristics and production as depicted by the blue line in Fig. 5.1. Since production from shale wells has become possible due to a significant amount of human intervention (in the form of completing long laterals with a large number of hydraulic fractures), many operators

Fig. 5.1 Conventional wisdom states that productivity in a well increases with reservoir quality

started asking a question that used not to be asked often and was always taken as the ground truth. The question was directed toward the impact of reservoir characteristics (rock quality) and its relationship with completion practices.

At the first glance it may seem that such question should be easy to answer. If the answer is not quite obvious from the operations (which one will quickly realize that it is not—please see Figs. 5.14 and 5.15 at the end of this chapter as examples where no trend is visible when raw data is plotted), then we can refer to our models for the answer. The procedure should not be very complicated. In our models, we can keep the completion and hydraulic fracturing characteristics constant and change the reservoir characteristics and observe its impact on production and then answer the above question. It sounds pretty simple and straightforward, until one realizes that such models (capable of realistically addressing questions such as this) do not exist. State of reservoir modeling in shale was discussed in Chap. 2.

In other words, the formulations that are currently used to model fluid flow (and therefore production) in shale, do not really represent what is happening, and therefore, scientists and engineers cannot fully trust the results generated by these models. This is true at multiple levels, including the modeling of the storage, the transport of the fluids, and the induced fractures.

In order to answer the question posed earlier for a given asset we will only refer to actual field measurements, or as we call them "Hard Data." There is no claim that the results shown here are general in nature. We recommend similar study to be applied to each field. Hard data is defined as field measurements such as inclination, azimuth, well logs (gamma ray, density, sonic, etc.), lateral and stage lengths, fluid type and amount, proppant type and amount, ISIP, breakdown and closure pressures, and corresponding injection rates.

As far as the reservoir characteristics are concerned, we use measurements such as net pay thickness, porosity, gas saturation, and TOC to define rock quality. Furthermore, we use pressure-corrected production as indicator of productivity. We use the Shale Analytics technology that was introduced earlier in this book (Chap. 3) for the analyses that are presented here.

5.2 Shale Formation Quality

Supervised Fuzzy Cluster Analysis is designed such that it allows engineers and geoscientists with domain expertise to define the shale (formation) quality. This is a simple but very important modification to the Fuzzy Cluster Analysis algorithm in order to accommodate the type of analysis that is presented here. Again, the objective of this analysis is to answer a specific question regarding the importance and the influence of reservoir quality on production from shale well sand how they differ from that of completion practices. Such analyses would not have been possible without making the required modification to the Fuzzy Cluster Analysis algorithm.

The modification to the original Fuzzy Cluster Analysis algorithm (to add the supervision to the algorithm) is based on a simple observation that allows us to impose certain domain expertise into our purely data-driven analysis. In other words, we attempt to address a common observation by engineers and geoscientist when they are exposed to the data-driven analytics. Since we do know certain underlying physics regarding the shale quality, we will guide (supervise) our analysis in such a way, so that it can identify to what degree reservoir characteristics of shale in a given well is similar to the known physics. For example, if rock (shale) qualities can be distinguished such that differences between "Good" and "Poor" rock (shale) can be clearly emphasized, then it would be the objective of this analysis to learn the degree of membership of the reservoir quality of each well in each of these rock (shale) quality clusters. In other words, it should be then possible to determine the relative degree of "goodness" of the reservoir characteristics for a given well.

Next is judging the quality of the rock (shale), based on measured parameters. Since calculation of reserves in shale still is an ongoing topic of research, in order to be on the safe side and make the results of this study acceptable by engineers and scientists of all persuasions, we will not use any formulation to calculate reserves (as a proxy for reservoir quality) in shale. Instead, we will try to identify characteristics that are acceptable by almost anyone that has any background in reserve calculation of any type of formation, including shale. The rules of distinction between "Good" and "Poor" rock (shale) qualities will be based on simple observations, such as the following:

Everything else being equal;

1. Formations with Thicker Net Pay should have more hydrocarbon reserves than formations with Thinner Net Pay.
2. Formations with higher Porosity should have more hydrocarbon reserves than formations with lower Porosity.
3. Formations with higher values of Hydrocarbon Saturation should have more hydrocarbon reserves than formations with lower values of Hydrocarbon Saturation.
4. Formations with higher TOC should have more hydrocarbon reserves than formations with lower TOC.

We will use the above four noncontroversial rules in order to define reservoir quality in shale. The measured data will be the foundation for classifications. It is important to note that as long as we are committed to work with actual field measurements, then we can only work with the "data we have" rather than the "data we wished we had." In this particular field, there were four reservoir characteristics available: Net Pay Thickness, Porosity, Gas Saturation, and TOC. In other cases that more data may be available (geo-mechanical characteristics), they can also be involved with defining the rock (quality) quality.

Referring to the rules identified above (and as demonstrated in Fig. 5.2) we define and impose (supervise) the locations of three cluster centers as "Good" (larger red circle), "Average" (larger green circle), and "Poor" (larger blue circle) shale reservoir qualities. In this way, each well with its given value for these four

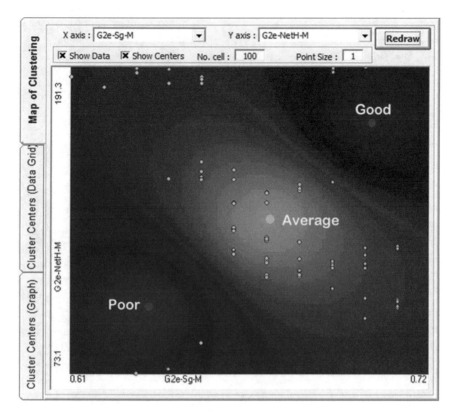

Fig. 5.2 Plot of net pay thickness versus gas saturation. The smaller *white circles* identify the location of measurements for each well. The location of *colored larger circles* identify our definition of "Good," "Average," and "Poor" rock qualities

parameters will acquire a membership in all three clusters. In other words, each well in this field is assigned a set of three memberships. The formation surrounding each well is "Good," "Average," and "Poor," each to a degree (please see Fig. 5.3).

Using this technique we have achieved two important objectives. Figures 5.2 and 5.3 both are the cross plots of Net Pay Thickness and Gas Saturation. Similar cross plots are generated for all the combinations of these four reservoir characteristics and the cluster centers for rock qualities "Good," "Average," and "Poor" are defined. It should be noted that based on these definitions, we now have a clear, and noncontroversial, definition for "Good," "Average," and "Poor" shale reservoir qualities. Furthermore, thanks to the Supervised Fuzzy Cluster Analysis algorithm we know to what degree (fuzzy membership function) each well is completed in which of these reservoir qualities. For example the well #1 identified in Fig. 5.3 (each small white circle represents reservoir characteristics measurements for a single well) has membership in all three fuzzy sets of "Good," "Average," and

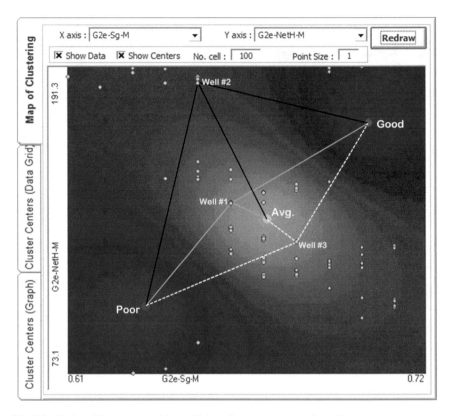

Fig. 5.3 Each well has membership in all three fuzzy sets (reservoir qualities)

"Poor," but each to a degree. As it is shown in this figure this particular well is represented by "Average" reservoir quality far more than by the other two clusters.

Figures 5.4 and 5.5 provide graphical as well as statistical information regarding the results of the Supervised Fuzzy Cluster Analysis. Figure 5.4 shows that cluster of wells with "Good" reservoir quality end up having a higher value of Net Pay Thickness, Porosity, Gas Saturation, and TOC than the wells completed in areas of the reservoir identified as "Average," and "Poor" reservoir characteristics. Furthermore, Fig. 5.5 shows that the statistics of reservoir characteristics for each of the clusters support our original intent of classifying wells based on the quality of the reservoir that they are completed in.

Statistics in Fig. 5.5 shows that there are 39 wells that have been completed in parts of the reservoir with dominant membership in the cluster of "Poor" reservoir quality, 127 wells have been completed in parts of the reservoir with dominant membership in the cluster of "Average" reservoir quality, and 55 wells have been completed in parts of the reservoir with dominant membership in the cluster of "Good" reservoir quality. Now that we have identified the degree of membership of each well in the relevant clusters we continue our analysis by trying to first identify

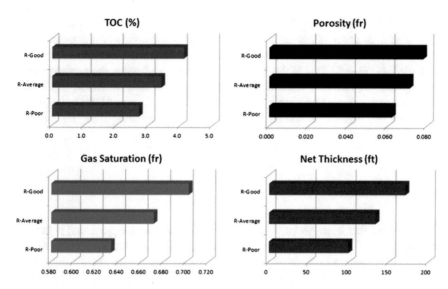

Fig. 5.4 Results of the imposition of the rules mentioned in this section on the field data in classification of the shale reservoir quality

Fig. 5.5 Statistics about each of the defined fuzzy clusters of reservoir qualities

Cluster Statistics

Clusters	1	2	3
Well Quality	R-Poor	R-Average	R-Good
No. of Wells	39	127	55
Avg. Entropy	0.45	0.31	0.47
Avg. Membership	0.61	0.73	0.58
TOC (%)	2.7	3.4	4.1
Porosity (fr)	0.062	0.071	0.078
Net Thickness (ft)	99	133	170
Gas Saturation (fr)	0.633	0.671	0.702

and compare (to one another) the production behavior of wells in these categories and then we will try to identify the completion parameters that are dominating a certain behavior in each category of the wells.

Before continuing, let us introduce the production indicator as the last parameter that needs to be calculated for each well in this analysis. The production indicator is the pressure-corrected, three months cumulative production of each well. In this section of the analysis, we explore the production behavior of the wells that belong to each of the categories (clusters). The interest is to learn if the wells that were classified as "Poor" wells, based on reservoir characteristics (considering their degree of membership in that cluster) have lower production than wells that have been identified as "Average" wells, and "Good" wells? Also do wells that have been identified as "Average" wells based on reservoir characteristics (considering their degree of membership in that cluster) have lower production than wells that have been identified as "Good," and higher production than wells that have been identified as "Poor"? This is the "conventional wisdom," when the productivity of wells positively correlate with their reservoir characteristics. In other words, does the "conventional wisdom" apply to "unconventional resources"? We hope that the results of the analysis presented in the next section can shed some light on this question.

5.3 Granularity

One last item that needs to be explained before the results are presented is the idea of granularity. Granularity is defined as the scale or level of detail present in a set of data or other phenomenon. When discovering or analyzing patterns in a dataset, the idea of granularity becomes important. It is hypothesized that a trend or a pattern is valid once it can tolerate (remain consistent) a certain level of granularity. In other words, a trend and/or a pattern would be acceptable if it can hold (remain the same) as the granularity increases, at least one level. Furthermore, classes, clusters, or groups that form trends and patterns need a certain level of population to be judged as acceptable. The size of the population makes them to be accepted as a trend or pattern rather than an anecdote. For example it is not reasonable to expect a single well to represent a class. This would be more anecdotal evidence than trend or pattern.

There are no widely acceptable values or numbers for the number of classes, for the granularity, or population in a class. We may judge them based on our experience in the field that we are applying them to. In this study acceptable trends and patterns are those that hold at least one level increase in granularity. Furthermore, we postulate the acceptable minimum number of wells (population) in a cluster or category to be eight wells (almost one pad).

5.4 Impact of Completion and Formation Parameters

The results of these analyses are discussed in two sections. In the first section (here) the strategy for performing the analysis and how this strategy is implemented, is covered. Furthermore, in this section detection of general trends in data that explains the interaction between reservoir quality and completion practices in different rock qualities are explained. In the second (next) section the focus will switch on to completion (design) parameters where the influence of different completion (design) parameters are examined.

5.4.1   Results of Pattern Recognition Analysis

The question that is being addressed in this chapter of the book is: "Formation or Completion; which one is controlling the production in shale wells"? To answer this question we designed the following strategy:

(a) We develop qualitative definitions for the reservoir characteristics.
(b) Each well is assigned a membership to each of the clusters of reservoir qualities (Good, Average, and Poor).
(c) Productivity of each well is calculated based on the applied membership of its reservoir quality (cluster). This means that 100 % of the production from a well that is completed in a section of the reservoir with 100 % "Poor" reservoir quality will be allocated to cluster of "Poor" wells, etc.
(d) Productivity of wells in each cluster of reservoir quality are averaged to represent the productivity of the cluster.

To make this analysis as comprehensive as possible, we implement the above strategy in several steps. In the first step, let us start the process by dividing the wells in this asset into only two categories of "Good" and "Poor" reservoir qualities. Figure 5.6 shows the results of this analysis. In this figure, the horizontal bar charts on the right show that all the 221 wells in this field have been completed in regions that are divided into "Good" and "Poor" reservoir qualities and that "Good" reservoir quality is consisted of thicker shale with better porosity and gas saturation and higher values of TOC.

Furthermore, this figure shows (vertical bar chart on left of Fig. 5.6) that, as expected, based on the "conventional wisdom" wells completed in the "Good" parts of the formation have higher productivity than wells completed in the "Poor" parts of the formation. Based on this figure the 221 wells in this field are divided into 39 and 182 for "Poor" and "Good," respectively. Now let us see whether this conclusion will hold once we increase the granularity of the analysis from two categories (clusters) to three categories.

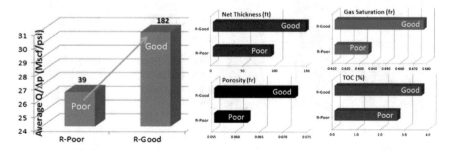

Fig. 5.6 Wells completed in areas with "Poor" reservoir quality have lower productivity than well completed in areas with "Good" reservoir quality

Figure 5.7 shows the results of this analysis when the granularity is increased from two to three categories of reservoir quality. In this figure the horizontal bar charts on the right clearly show the values of reservoir characteristics representing the classifications of "Good," "Average," and "Poor." Of the 221 wells in this field, the 39 wells in the "Poor" cluster remain in this category as before, while the remaining wells are divided into 127 and 55 wells in the clusters "Average," and "Good," respectively. The vertical bar chart on the left of Fig. 5.7 shows that the productivity no longer follows the expected trend. In other words, the wells completed in areas with "Average" reservoir quality have produced better than the wells in the areas with "Good" reservoir qualities. This is an unexpected result. But before we make any conclusions, we first have to make sure that the trends that are observed in Fig. 5.7 can withstand the scrutiny of increase in granularity.

To do this, the granularity of each section of the bar graphs shown in Fig. 5.7 is increased from two to three categories. In other words, concentrating on the "Poor" to "Average" part of the field, we will increase the granularity of this section from two to three clusters and then switching concentration to the "Average" to "Good" part of the field, we will increase the granularity of this section also from two to three clusters. Therefore in general, one can argue that we have increased the granularity of the analysis for the entire field from three to five or even six clusters (depending how one would look at this, since there will be an overlap of 34 wells between the wells in the best cluster of one classification and the wells in the worst cluster of the next classification, as will be explained in the next few paragraphs).

In Fig. 5.8 the concentration is shifted only to the wells that have been completed in "Poor" to "Average" part of the field. We will call these parts of the reservoir the "Low Quality Shale—LQS." In Fig. 5.9 the concentration shifts to the wells that have been completed in "Average" to "Good" part of the field. We will call these parts of the reservoir the "High Quality Shale—HQS." These figures show that as the granularity in each of these types of reservoirs increases from two to three categories, the trend that was first observed in Fig. 5.7 holds. Furthermore, in the LQS, Fig. 5.8, the horizontal (reservoir quality) bar charts are in agreement

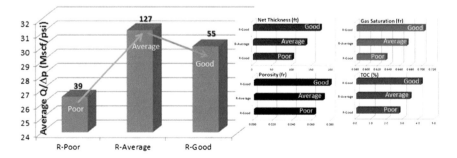

Fig. 5.7 Wells completed in areas with "Poor" reservoir quality have lower productivity than well completed in areas with "Average" reservoir quality and these wells have higher productivity than well completed in areas with "Good" reservoir quality. An unexpected result

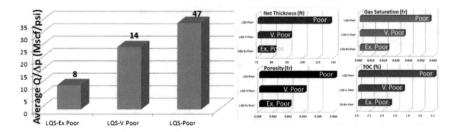

Fig. 5.8 For low quality shale (LQS) the productivity trend matches that of reservoir quality

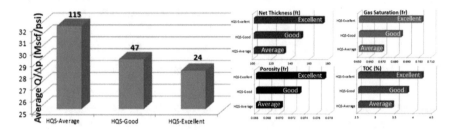

Fig. 5.9 For high quality shale (HQS) the productivity trend does not match that of reservoir quality

with the vertical (productivity) bar chart, while in the HQS, Fig. 5.9, the opposite is true. The trends in Figs. 5.8 and 5.9 mirror the trends shown in Fig. 5.7 (the left vertical gray, bar chart) but at higher granularity.

The only logical explanation of these patterns is that while the "conventional wisdom" seem to hold for the Lower Quality Shale (LQS), it does not necessarily hold for the Higher Quality Shale (HQS). In other words, while reservoir quality seems to dominate production behavior in the LQS, it takes a backseat to other factors (such as completion practices) in the HQS. In the LQS, completion does not do much more than allowing the rock to behave in a matter that is expected of it, and completion simply provides the means for the rock to express itself (in terms of production) as it should. In the LQS operators get more production from wells located in the better parts of the reservoir, and as long as the completion practices are within acceptable industry range (nothing special) they should expect to get acceptable return on their completion investment.

However, in the HQS, the role and impact of the completion practices becomes much more pronounced. In the HQS if the completion practices are not carefully examined and designed based on sound engineering judgments and detail scientific studies, they can actually hinder the capabilities of shale in producing all that it is capable of. Figure 5.9 provides the insight that is the source of this reasoning. This figure shows that as far as the reservoir characteristics are concerned, wells located in "Good," to "Excellent" parts of the reservoir are less productive than those located in the "Average" parts of the reservoir. Therefore, if the reservoir quality is not driving the production, then what is? The only other potential culprits are completion and well construction practices.

Therefore, for the HQS they (completion and well construction practices) must be influencing the production such that they are overshadowing the influence of reservoir characteristics. Please note that these are not anecdotal observations about one or two wells. This is a pattern that includes 221 wells. Figure 5.9 suggests that this operator is not getting the type of productivity from its shale wells that it should, and therefore the completion practices and design can and should be improved. But How? Which completion parameters are actually controlling the productivity of these wells? And how should they be modified in order to improve productivity? These are the questions that will be addressed in the next section of this chapter, as we continue the implementation of Shale Analytics.

5.4.2 Influence of Completion Parameters

Now that we have established that the influence of completion practices on production in shale wells is a function of reservoir quality, the focus will be switched to specific completion characteristics in order to determine their influence on the productivity of the shale wells. Following is a detail explanation of how this is accomplished. Each well is qualitatively classified based on the definitions of reservoir characteristics. Then the degree of membership of wells in each of the clusters is used as an indicator to calculate its production, as well as each of the completion variables and averaged for all the wells in that cluster. If the resulting productivity and completion variable show similar trends then the conclusions are made accordingly. Figures 5.10, 5.11, 5.12 and 5.13 show this type of analyses.

Let us examine Fig. 5.10 to clarify this algorithm. This figure shows the analysis for the Low Quality Shale (LQS). The three reservoir qualities have been identified as "Extremely Poor," "Very Poor," and "Poor" and the corresponding reservoir characteristics are shown using horizontal bar charts on the top-right of the Fig. 5.10 (this is the same bar chart better shown in Fig. 5.8, on the right). These charts clearly show that the average Net Thickness, Porosity, Gas Saturation, and TOC of the "Extremely Poor" shale is less than those of "Very Poor" shale and those of "Very Poor" are less than of those for "Poor" shale. In other words, the classification seems to be quite justified.

When the productivity of the wells that are classified as above, are plotted (top left vertical, gray bar chart in Fig. 5.10), one can observe that as expected, productivity of the wells producing from "Poor" reservoir rock is higher than the wells producing from "Very Poor" reservoir rock, and so forth. When the membership of the wells based on their reservoir quality is used to calculate their share of the completion attribute and then plot them accordingly, one can see that for example in the case of "Total Number of Stages" (brown vertical bar chart on the middle right side of Fig. 5.10) the trend is similar to that of production. In this case we make two

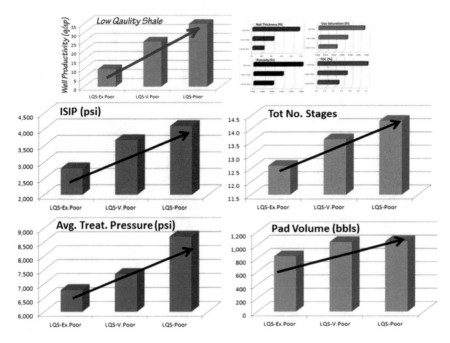

Fig. 5.10 Impact of completion characteristics on production in low quality shale; Trend similar to production

conclusions for the LQS: (a) this particular attribute, "Total Number of Stages," is a dominant (monotonic) attribute (since it has a dominant non-changing trend) and (b) based on the direction of the trend, higher values of "Total Number of Stages" cause better productivity. Similar conclusions can be made for attributes such as "Average Treatment Pressure" (the wells completed in the [relatively] better quality shale [within the LQS general category] should be treated with higher average injection pressure) and "Amount of Pad Volume" (wells completed in the better [relatively] quality shale [within the LQS general category] should be treated with larger amounts of pad volume) based on the two vertical bar charts at the bottom of Fig. 5.10.

Figure 5.11 shows the trend analysis (LQS) for completion parameters that seem to have a dominant but opposite impact on productivity. Using the same logic presented in the previous paragraph, conclusions can be made on three other parameters. For amount of "Proppant per Stage," it can be mentioned that based on the orange vertical bar chart on the bottom of Fig. 5.11 (when compared with the trend of productivity) the data suggest that for the LQS, higher values of "Proppant per Stage" seem not to be an appropriate design consideration. This figure shows that shale wells completed in the "Extremely Poor" parts of the reservoir cannot produce better even when larger amounts of "Proppant per Stage" are used during their completion.

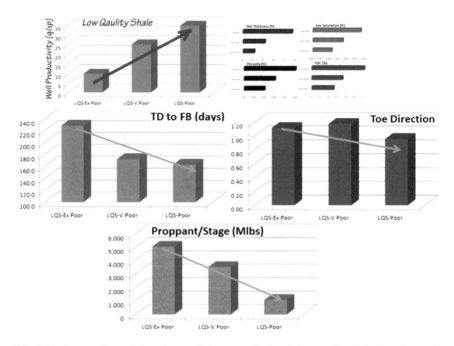

Fig. 5.11 Impact of completion characteristics on production in low quality shale; Trend opposite to production

Similar conclusions can be made regarding the "Soak Time." The analysis show that in this particular field, for the LQS, the shorter soak time (number of days between completing the well and flowing it back) seem to be beneficial (please note the soak time in this field is generally high).

To analyze the completion parameters in the High Quality Shale in this filed Figs. 5.12 and 5.13 are presented. The logic is similar to those presented for the LQS. The top-left (vertical, gray) bar chart in this figure shows an unintuitive plot. Here it is seen that as the quality of the rock gets better (x axis) the productivity decreases (the bars). This is exactly the opposite of what is expected. So what is causing this? The bottom left, brown, vertical bar chart shows that in the wells that have been analyzed in this field the "Total Number of Stages" have become less as the quality of the rock gets better.

A clear positive correlation exists between lower productivity (in the shale wells completed in better quality rock) and "Total Number of Stages." This may explain (at least partially) the lower productivity of the wells that were supposed to produce better. Similar conclusions can be made for completion parameters such as "Proppant Concentration" and "Pad Volume."

Completion parameters with opposite (negative) correlation for the HQS are "Length of Each Stage" and "Percent of Fine Proppant." Figure 5.13 suggest that in the case of both of these parameters lower productivity is directly and positively

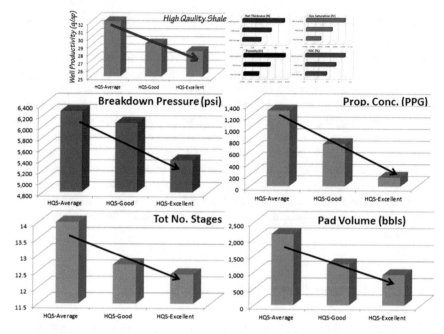

Fig. 5.12 Impact of completion characteristics on production in high quality shale; Trend similar to production

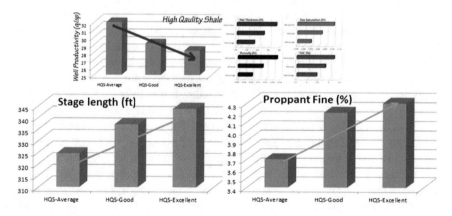

Fig. 5.13 Impact of completion characteristics on production in high quality shale; Trend opposite to production

correlated with higher values. In other words, in the shale wells that have been completed in the better parts of the reservoir, the productivity has suffered mainly because these wells have been completed with larger stage lengths and higher values of fine sized proppant.

5.4.3 *Important Notes on the Results and Discussion*

The reader may have noticed that throughout the above analyses we have avoided using numbers. When the LQS and the HQS are mentioned no numbers are presented (although one can read it on their own from the axes). There is a good reason for this. As engineers, we have been conditioned to continuously deal with numbers and associate everything with a scale. That is fine and appropriate with many of the analyses that we perform day in and day out as engineers and geoscientists. However, when it comes to pattern analyses almost all of them are "fit-for-purpose." Specifically, when it comes to shale, we would like to warn against using the results presented here and generalizing them in any shape or form to any other shale basin and/or even other fields in the same basin. If you have enough data, then we suggest to perform similar analyses and make the appropriate conclusions.

5.5 Chapter Conclusion and Closing Remarks

The application of Shale Analytics that was presented in this chapter provides the type of insight that is required in order to dig deeper into the completion practices in the shale plays. It was demonstrated that pattern recognition technology that is an integral part of Shale Analytics can shed important light on the influence of design parameters in shale wells productivity. This technology can distinguish between the impact of rock quality and those of completion practices on well productivity. It was shown that completion design is not as important in wells completed in Low Quality Shale (LQS) as they are in wells completed in High Quality Shale (HQS).

While "One-size-fit-all" design philosophy may be sufficient for LQS, it certainly is not (and short sells) the HQS. The author firmly believes that by changing and optimizing the design of the completion and hydraulic fracturing in shale, much more can be expected from this prolific resource. The question that was asked in the beginning of this chapter was "Formation or Completion; which one is controlling the productivity"? It was demonstrated that this is not an easy question to answer and the answer is different for every field.[1] However, for the field that was the subject of this chapter, it was demonstrated that Shale Analytics is capable of using facts (field measurements) in order to provide answers to help the operators in their quest to optimize production from shale.

An overall look at the analyses presented in this chapter results in the conclusion that "we already knew many of the conclusions that we represented in this chapter."

---

[1]If similar analyses are performed on multiple fields in multiple assets, then we may be able to produce some general conclusions. At this point, it is safe to conclude theses are field dependent analyses and should not be generalized.

Why should one bother with all these details in order to reach the conclusions that are so intuitive (such as more stages are better)?

Well, the more important question should be "do we need to learn from the measured data (facts from the field)"? One of the main conclusions of this chapter that will be the main theme throughout this book is that in both low and high quality shale, it is better to have more number of stages. This is quite intuitive and most reservoir and production engineers will tell you that they already knew this. This holds true since in shale you only produce where you make contact with the rock, and although not all your hydraulic fractures are successful, by increasing the number of hydraulic fractures (stages) one increases the chances of success. This is indeed a true statement. However, the more important question is "can one make the same conclusion by looking at the actual raw data from the field"?

Figure 5.14 is the cross plot of best 12 months of cumulative production versus total number of stages that was used in the analyses. Can anyone make the conclusion that "in this field, for better productivity, regardless of the rock quality, it is better to have a higher number of stages" by looking at this plot? Once it is demonstrated (by the analysis presented in this chapter) that such a reasonably intuitive conclusion can indeed be made from the actual data, regardless of its seemingly chaotic behavior (as shown in Fig. 5.14), then one may feel more

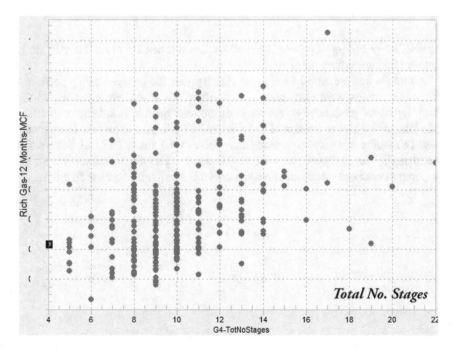

Fig. 5.14 Cross plot of 12 months cumulative production versus total number of stages, on a per well basis

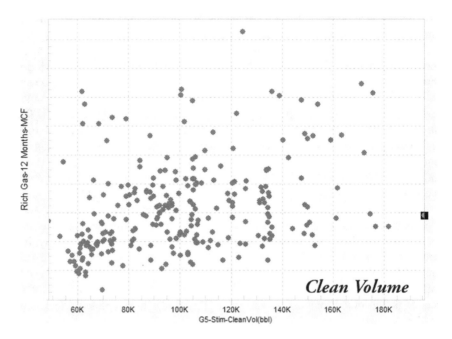

Fig. 5.15 Cross plot of 12 months cumulative production versus total amount of clean (pad) volume, on a per well basis

confidence regarding other conclusions that are reached by these analyses (technology) that are not so intuitive.

Although not presented here in its entirety (for the confidentiality purposes), other conclusions were also made as a result of this study. For example "in this field, for better productivity, regardless of the rock quality, it is better to start the frac jobs with a higher amount of pad volume." Figure 5.15 shows the cross plot of best 12 months of cumulative production versus total amount of pad volume that was used in the analyses provided in this article. This is an unintuitive but yet important conclusion that can serve the operator in optimizing their frac jobs.

Chapter 6
Synthetic Geomechanical Logs

In recent years, predicting reservoir's behavior as a function of changes in field stresses has become a focus in the E&P industry, with an increased interest to unlock production from unconventional resources. Geomechanics integrates solid and fluid mechanics, physics, and engineering geology to determine the rock and fluid flow responses to stress changes due to drilling, production, fracturing, and EOR. When there is a change of stress in a rock, it deforms and alters its volume and geometry and subsequently the paths of fluid flow. Hence, every stage in the reservoir life cycle benefits from the accurate determination and evaluation of geomechanical properties.

6.1 Geomechanical Properties of Rocks

Failure to accurately understand the geomechanics will have consequences especially for shale assets. Excessive mud loss, wellbore instability, casing compression and shearing, sand production, and shale horizontal well's quick production decline are some of the results of ignoring stress changes. Thus, drilling design, well bore stability, production planning, and more importantly hydraulic fracturing design are highly dependent on the geomechanical properties and models of reservoir rocks. Geomechanical properties are normally measured by laboratory tests on cores or more recently obtained using some advanced geomechanical well logs. This section describes the geomechanical properties of interest to shale operations. These include, Minimum Horizontal Stress, Young's Modulus, Shear Modulus, Bulk Modulus, and Poisson's Ratio.

This Chapter is Co-Authored by: Dr. Mohammad Omidvar Eshkalak, University of Texas.

© Springer International Publishing AG 2017
S.D. Mohaghegh, *Shale Analytics*, DOI 10.1007/978-3-319-48753-3_6

6.1.1 Minimum Horizontal Stress

Reservoirs are confined and under three principal stresses, vertical, maximum, and minimum Horizontal stresses. The magnitude and direction of the principal stresses are important because they control the pressure required to create and propagate a fracture, the shape, vertical extent, and its direction. A hydraulic fracture will propagate perpendicular to the minimum horizontal stress. Figure 6.1 shows the Minimum Horizontal Stress.

6.1.2 Shear Modulus

Shear modulus is concerned with the deformation of a rock when it experiences a force parallel to one of its surfaces. Figure 6.2 shows the forces applied and its appropriate formulation. The unit for Shear modulus is Pascal in SI.

6.1.3 Bulk Modulus

Bulk modulus of rocks measures rock resistance to uniform compression. In brief, bulk modulus is exactly the reciprocal of bulk compressibility. Figure 6.3 shows the bulk modulus definition schematically. Bulk modulus base unit in SI is Pascal.

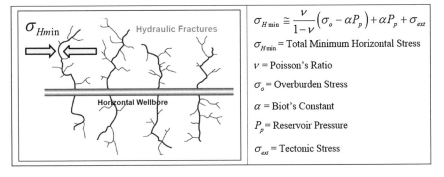

$$\sigma_{H\min} \cong \frac{\nu}{1-\nu}\left(\sigma_o - \alpha P_p\right) + \alpha P_p + \sigma_{ext}$$

$\sigma_{H\min}$ = Total Minimum Horizontal Stress

ν = Poisson's Ratio

σ_o = Overburden Stress

α = Biot's Constant

P_p = Reservoir Pressure

σ_{ext} = Tectonic Stress

Fig. 6.1 Total minimum horizontal stress

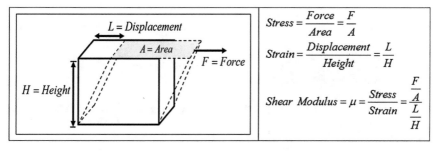

Fig. 6.2 Shear modulus

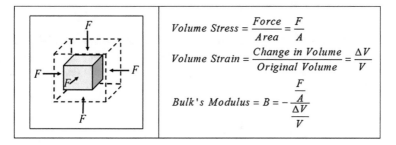

Fig. 6.3 Bulk modulus

6.1.4   Young's Modulus

Young's modulus is defined as the ratio of the stress along an axis over the strain along that axis. Figure 6.4 shows the schematic of the young modulus. Young's modulus has units of pressure, Pascal.

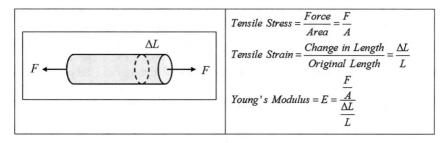

Fig. 6.4 Young's modulus

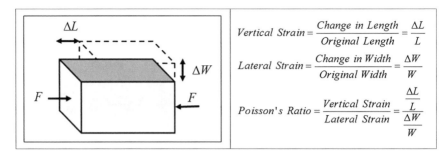

Fig. 6.5 Poisson's ratio

6.1.5 *Poisson's Ratio*

Poisson's ratio is the negative ratio of transverse to axial strain. In brief, Poisson's ratio represents the amount that the sides bulge out when the top is compressed. Figure 6.5 depicts the Poisson's ratio schematically. Poisson's ratio is unit less.

6.2 Geomechanical Well Logs

Well logging is the practice of making a detailed record of the formations penetrated by a borehole. The oil and gas industry uses logging to obtain a continuous record of a formation's rock properties. Well logging is used for almost a century in the petroleum industry as an essential tool for determination and classification of potential production formations in oil and gas reservoirs. Among different well logs, geomechanical logs are considered as recent advanced practice within service companies. The costs of running these logs are reported to be higher than any other type of conventional logs, especially for horizontal shale wells. These geomechanical well logs include Minimum Horizontal Stress, Poisson's ratio, Young's, Bulk, and Shear Moduli.

While rig rental cost has increased during past years, the costs of running different logs have also increased. The need for fast, efficient, and accurate log acquisition has become more prevalent these days. Other methods such as performing lab experiments on cores are costly. Despite the popularity of the geomechanical logs in recent years, companies do not always run all the logs for all available wells within an asset. Therefore, methods that are capable of generating accurate synthetic well logs (especially synthetic geomechanical well logs) can become quite attractive solutions.

6.3   Synthetic Model Development

Marcellus Shale, a largely untapped reserve, is estimated to contain 500 trillion cubic feet (Tcf) of natural gas. Its proximity to the high-demand markets along the East Coast of the United States makes it an attractive target for energy development. Thickness of Marcellus shale ranges from 890 ft in New Jersey, to 40 ft in West Virginia. The portion of the Marcellus shale that is of interest in this chapter is located in southern Pennsylvania and northern West Virginia. The portion of the asset being used in the analysis that is presented in this chapter includes 82 horizontal wells. Figure 6.6 shows a map of these wells schematically.

In this chapter, we show how Shale Analytics is used to generate synthetic geomechanical logs (Including logs for Total Minimum Horizontal Stress, Poisson's Ratio, Bulk, Young's, and Shear Modulus) for 50 wells (out of the 80 wells shown in Fig. 6.6) that do not have such logs. Geomechanical properties are key components that directly control production in the Marcellus shale. Since hydraulic fracturing as well as the consequent production of hydrocarbon can significantly alter the stress profiles and geomechanical properties throughout the reservoir, it is important to understand and generate these parameters for the entire Marcellus Shale reservoir. The synthetic geomechanical logs computed based upon their available conventional logs (such as Gamma Ray and Density Porosity). Table 6.1 shows the typical values of geomechanical properties of Marcellus Shale in the area being investigated in this chapter.

The approach to generate synthetic geomechanical well logs for this asset is to incorporate conventional logs such as gamma ray and bulk density that are relatively inexpensive and commonly available for the assets. The well logs for 80

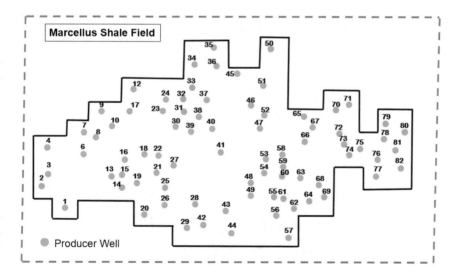

Fig. 6.6 Location of the Marcellus shale wells used in the study presented in this chapter

Table 6.1 Typical geomechanical properties of Marcellus Shale

Parameter	Average value	Unit
Total minimum horizontal stress	0.8	Psi/ft
Bulk modulus	4	Mpsi
Shear modulus	2.5	Mpsi
Young modulus	3.5	Mpsi
Poisson's ratio	0.15	Fraction

wells in the asset are categorized into two main datasets. The first group of 30 wells has both conventional as well as geomechanical well logs provided by a service company through running the logs. The second group of 50 wells only has conventional well logs. Relative locations of these two groups of wells are shown in Fig. 6.7.

The second group of wells (that includes 50 wells with only conventional well logs), are further divided into three subsets of 30, 10, and 10 wells. These subsets are defined based on the available conventional logs such as Sonic Porosity, Gamma Ray, and Bulk Density. The first subset (30 wells) includes all conventional well logs (Sonic Porosity, Gamma Ray, and Bulk Density). The second subset of wells containing 10 wells are missing Sonic Porosity (include only Gamma Ray and Bulk Density), and finally the last subset includes 10 wells that only have Gamma Ray (missing Sonic Porosity, Bulk Density). It must be noted that the available well logs for all 80 wells cover the shale pay-zone as well as the non-shale section of the reservoir. Table 6.2 shows the description of the well log availability for all the 80 well in this asset.

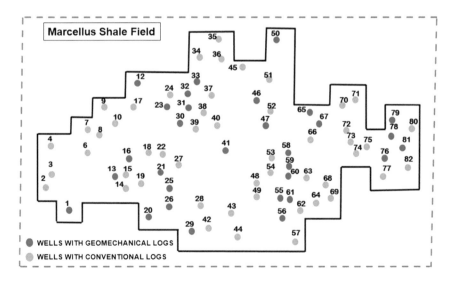

Fig. 6.7 Marcellus Shale asset map showing both groups of wells

Table 6.2 Description of well log availability for the two groups of wells

Group category	Number of wells		Geomechanical logs	Conventional logs		
				Sonic porosity	Bulk density	Gamma ray
1st	30		Yes	Yes	Yes	Yes
2nd	50	30	No	Yes	Yes	Yes
		10	No	No	Yes	Yes
		10	No	No	No	Yes

6.3.1 Synthetic Log Development Strategy

The very first step in the strategy of how to solve this problem is to make sure that the final solution would be robust and acceptable. In order to accommodate this important task the solution must be validated against wells that have geomechanical well logs, but have not been used to develop the solution. We call these wells "Blind Wells" (they may also be referred to as validation wells). Five blind wells are selected and removed from the group of 80 wells from the start of the project. These "Blind Wells" are selected from the first group of 30 wells that have all the conventional and the geomechanical well logs. Figure 6.8 shows the location of the "Blind Wells" in the field. Therefore, the quality of the solution (the model that would generate the synthetic geomechanical logs) will be judged as a function of its performance against the "Blind Wells."

Given the fact that twenty (20) wells are also missing conventional well logs (10 wells are missing Bulk Density logs and 20 wells are missing Sonic logs,

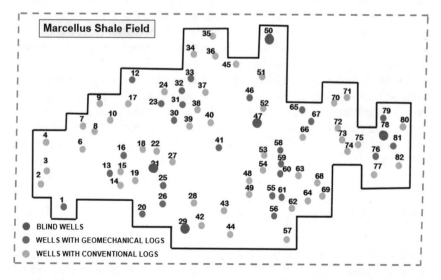

Fig. 6.8 Wells in the field used for training and calibration of the models and the "Blind Wells"

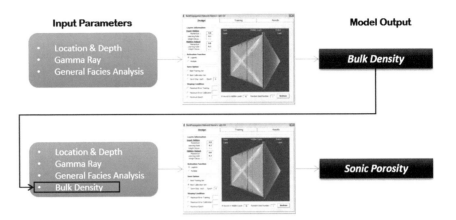

Fig. 6.9 The process developed for generating synthetic conventional well logs

Table 6.2), the strategy for this project will include an extra step to generate synthetic conventional well logs for these wells so that they can participate in the generation of the synthetic geomechanical well logs. Figure 6.9 demonstrates how the models for generating synthetic conventional well logs were generated. The process of generating synthetic conventional well logs were performed in the past on several fields [64–66].

As shown in Table 6.2 there are 70 wells (65 wells used for training and calibration and 5 blind wells) that can be used to develop a model for the generation of the synthetic Bulk Density Logs and there are 60 wells that can be used to develop a model for the generation of the synthetic Sonic Porosity Logs as shown in Fig. 6.9. Once the process shown in Fig. 6.9 is completed, we will be able to use the model developed for the generation of the synthetic geomechanical well logs to generate synthetic geomechanical well logs for all the 50 wells in the second category of the well as shown in Table 6.2. The model developed to generate synthetic geomechanical well logs is shown in Fig. 6.10.

The development process of the data-driven models for the generation of the synthetic well logs (either conventional or geomechanical well logs) includes the process of training, calibration, and validation of the data-driven model using the dataset generated for this purpose from all the available well logs. This is prior to the application (deployment) of the data-driven model to the blind wells for final validation.

6.3.2   Results of the Synthetic Logs

The main database contains the well name, the depth, the well coordinates, the values for gamma ray, bulk density, sonic porosity, bulk modulus, shear modulus,

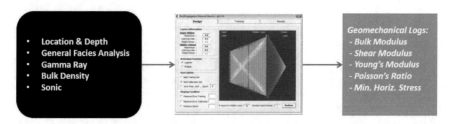

Fig. 6.10 The process developed for generating synthetic geomechanical well logs

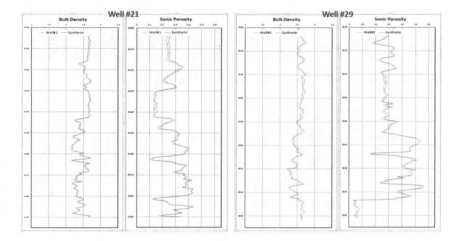

Fig. 6.11 Results of synthetic conventional models for the Blind Well #21 and #29

Young's modulus, Poisson's ratio, and the total minimum horizontal stress for each well. Please note that wells that lack certain well logs as clarified in Table 6.2, will have empty spaces for such logs in the database and the objective is to complete (fully populate) the database using the models that will be generated with synthetic logs.

Using the models shown in Fig. 6.9 synthetic conventional logs are generated for the five blind wells shown in Fig. 6.8. Comparison between actual well logs (measured by the service companies in the field) and the synthetic conventional well logs are shown in Figs. 6.11 and 6.12. It is clear from these figures that the synthetic conventional well logs are quite accurate.

Upon populating the database with the results generated from the models shown in Fig. 6.9 the natural next step is to develop (train, calibrate, and validate as covered in the previous chapters) the synthetic geomechanical well logs. Using the process demonstrated in Fig. 6.10 synthetic geomechanical well logs for Bulk, Shear, and Young's Moduli, Poisson's Ratio, and Minimum Horizontal Stress were

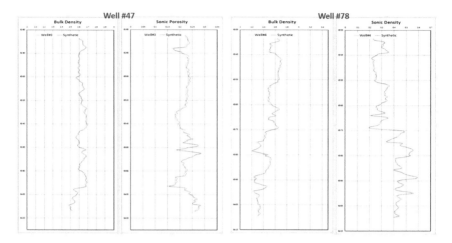

Fig. 6.12 Results of synthetic conventional models for the Blind Well #47 and #78

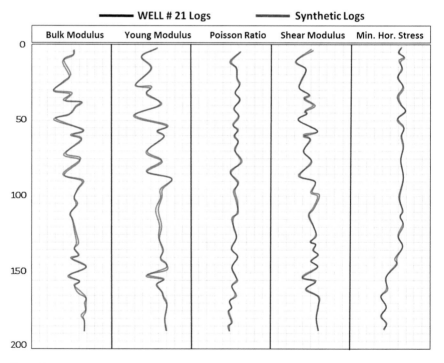

Fig. 6.13 Results of generating synthetic geomechanical well logs for the blind well #21

generated. The results of the blind test (applying the data-driven models for the generation of synthetic well logs to the five blind wells) are shown in Figs. 6.13, 6.14, 6.15, 6.16, and 6.17.

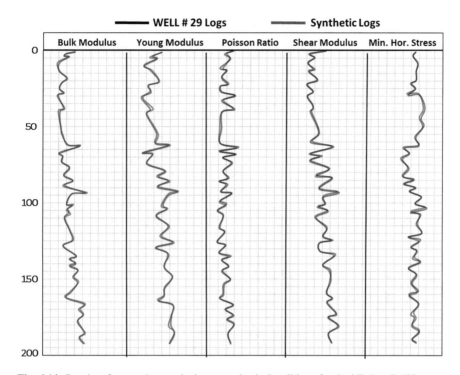

Fig. 6.14 Results of generating synthetic geomechanical well logs for the blind well #29

Results for blind wells #21, #29, #47, and #78 (Figs. 6.13, 6.14, 6.15, and 6.16) show how accurate the data-driven model can be in generating synthetic geomechanical well logs. However, results of the blind well #50 (Fig. 6.17) leaves much to be desired. In other words, the data-driven model completely fails to generate accurate synthetic geomechanical well logs for this well, but why? It makes no sense that the model performs so accurately for the other four wells but do such a poor job in the case of well #50. The immediate reaction to such a result for those that are familiar with how machine learning algorithms work is that the characteristics of this well falls outside of the ranges that were used to train this model. This is a legitimate conclusion, if and only if the premise (characteristics of this well falls outside of the ranges that were used to train this model) would be true (Figs. 6.14 and 6.15).

There is an easy way to test the above hypothesis. Since this is blind well, by definition, we have access to all its conventional and geomechanical well logs. Therefore, we can plot the ranges of all well logs and see if the well logs from well #50 falls outside of the ranges that were used to train the data-driven model.

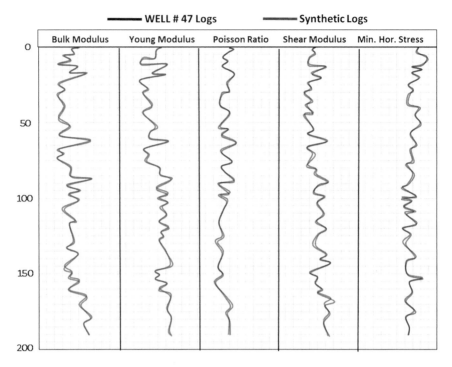

Fig. 6.15 Results of generating synthetic geomechanical well logs for the blind well #47

Figure 6.18 shows the result of checking this hypothesis. In this figure there are eight plots. In each of the plots the ranges of all the wells that were used to train the data-driven model are shown in a set of bar charts and the corresponding well log from the well #50 is superimposed (as a curve) on top of the bar charts. Interestingly enough seven out of the eight plots show that the well logs from well #50 is well within the range of the well logs that were used to train the data-driven model for the generation of the synthetic logs. The only well log that does not follow this trend is the Sonic Porosity log (the bottom plot to the far right).

Well, when Bulk, Shear, and Young's Moduli, as well as Poisson's Ratio, Minimum Horizontal Stress, Gamma Ray, and Bulk Density of a well falls within the range of all other wells, there seem to be no logical explanation for why the Sonic Porosity well log for this well should fall outside of the range? The only logical explanation may be that when the Sonic Porosity for this well was being

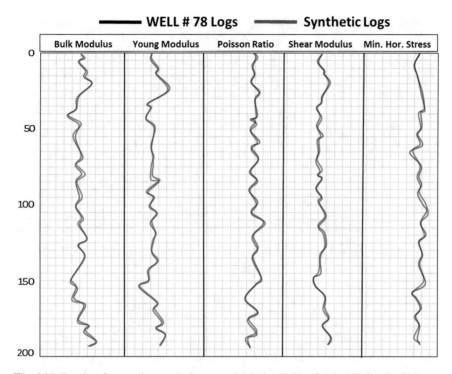

Fig. 6.16 Results of generating synthetic geomechanical well logs for the blind well #78

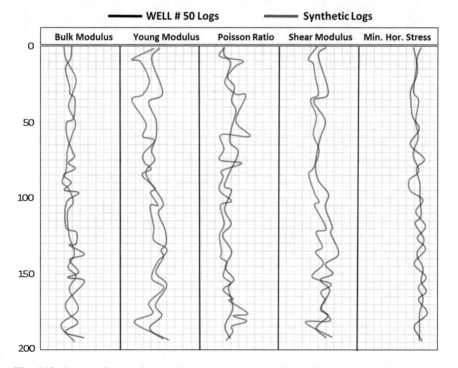

Fig. 6.17 Results of generating synthetic geomechanical well logs for the blind well #50

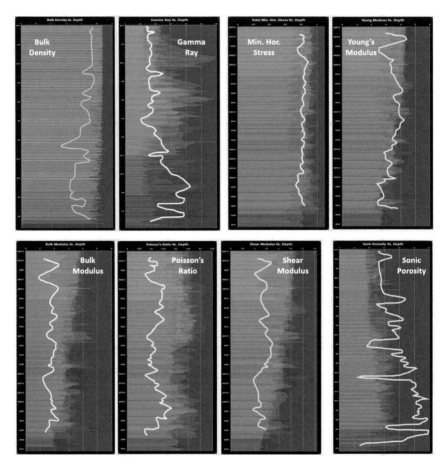

Fig. 6.18 Comparing the well log ranges of well #50 with the ranges of the wells used to train the data-driven model

measured in the field, the service company's instrument failed for some reasons and the measurements for Sonic Porosity were not recorded correctly. Of course, if this line of reasoning (or one may say this hypothesis) is correct, we should be able to prove it.

Let us assume that (a) the aforementioned hypothesis is correct, meaning that there had been a flaw in the measurement or the recording of the measurement of the Sonic Porosity log for Well #50, and (b) the data-driven synthetics well log

Fig. 6.19 Synthetic versus actual (measured) Sonic Porosity log for Well #50

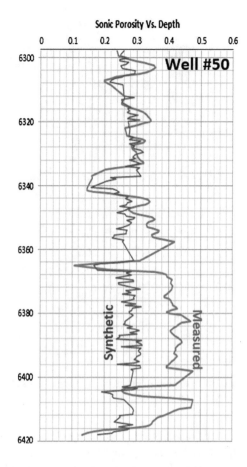

generation model developed in this study was actually an accurate model. If both of the assumptions are correct, then by substituting the actual but flawed (measured) Sonic Porosity log for Well #50 with the synthetic Sonic Porosity log for this well, and using it as input to the synthetic geomechanical well log model the problem should be solved.

In order to test this hypothesis, let us first plot the actual (measured) Sonic Porosity log for Well #50 against the synthetic Sonic Porosity log for the same well. This comparison is demonstrated in Fig. 6.19. As expected, there are marked differences between these two logs. The synthetic Sonic Porosity log for Well #50 looks nothing like the actual or measured Sonic Porosity log for this well. When the synthetic Sonic Porosity log for Well #50 is used to replace the actual (measured) sonic log as the input (to the data-driven model) to generate synthetic geomechanical well logs for this well, the results are much different, as shown in Fig. 6.20.

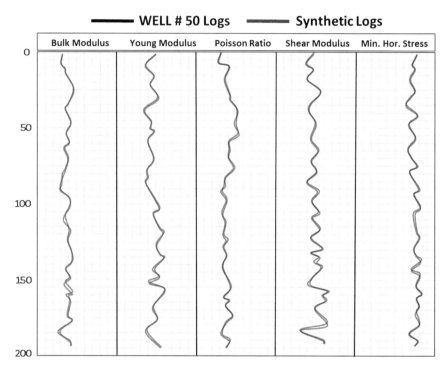

Fig. 6.20 Results of generating synthetic geomechanical well logs for the blind well #50 (using corrected Sonic Porosity Log)

This figure proves that the hypothesis regarding the measured Sonic Porosity log for Well #50 being inaccurate and must be thrown out, is a solid and acceptable hypothesis. This technique can be used in order to quality control the well logs from any field to make sure that there had been no equipment failure during the well logging in a given field.

6.4 Post-Modeling Analysis

Once the data-driven models for the generations of the synthetic geomechanical well logs are trained, calibrated, and properly validated using the five blind wells as shown in the previous sections, they are used to complete the database. Upon completion of the database and populating it with all relevant data, we have now completed the Table 6.2 and have "Yes" instead of "No" in all columns, meaning that we now have Bulk, Shear, and Young's Moduli, as well as Poisson's Ratio, and Minimum Horizontal Stress well logs for all the 80 well s in our database.

As it was mentioned at the beginning of this chapter, geomechanical properties of a shale reservoir is used to design effective hydraulic fractures. Figure 6.21

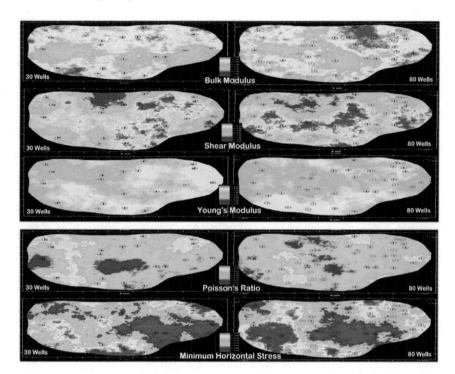

Fig. 6.21 Volumetric distribution of geomechanical properties in the Marcellus shale asset before and after using the data-driven model for generating synthetic geomechanical well logs

clearly shows the difference between volumetric distributions of the geomechanical properties in this asset when more wells (80 wells instead of 30 wells) are used to generate such properties.

In Fig. 6.21 the volumetric distributions of the geomechanical properties shown on the right hand side of the figure are generated by geostatistical techniques using 80 well logs as control points for distribution while the volumetric distributions of the geomechanical properties shown on the left hand side of the figure are generated by geostatistical techniques using 30 well logs as control points.

The difference between these two distributions for every single geomechanical property is quite obvious. It is a fact that in geostatistics, as the number of control points increases, so does the accuracy of the distribution. Therefore, there are far more information in the volumetric distributions of the geomechanical properties shown on the right hand side of the Fig. 6.21 than there is on the ones on the left hand side distributions. This can contribute significantly to better hydraulic fracture designs and result in better propagation of the induced fractures in the reservoir that would contact larger amounts of the reservoir volume and consequently produce much higher amounts of hydrocarbon.

Chapter 7
Extending the Utility of Decline Curve Analysis

Decline Curve Analysis (DCA) is one of the most utilized and probably the most simple (the least sophisticated) technique in petroleum engineering. DCA is a curve fitting technique that is used to forecast well's production and to Estimate its Ultimate Recovery (EUR). During DCA a predefined function (most of the time it is either hyperbolic, exponential, or an extension, combination, or modified version of such forms of equations) is used to fit the existing production data and then the function is extended in time to estimate the production in the future and the ultimate recovery of the well.

Before delving into the application of DCA in shale, authors of this chapter would like to put forward their view of the applicability of this technology to wells producing from shale formations. As it will become clear in the remainder of this chapter, we do not believe that DCA is a technically useful methodology for estimating ultimate recovery from shale wells. Nevertheless, given the fact that this technology is extensively used in the industry, this chapter attempts to demonstrate how Shale Analytics can increase the utility of this technology and extend its use by shedding some light on the (intended or unintended) consequences of using different forms of DCA to estimate ultimate recovery of shale wells.

7.1 Decline Curve Analysis and Its Use in Shale

When it comes to hydrocarbon production from shale, there is a major flaw in the application of DCA, whether it be the ARP's original formulation or other flavors that have emerged in the recent years [18–20]. Limitations associated with DCA are well known and have been discussed comprehensively in the literature. Nevertheless, most

---

This Chapter is Co-Authored by: Maher J. Alabboodi, West Virginia University & Faegheh Javadi, Mountaineer Keystone.

© Springer International Publishing AG 2017   127
S.D. Mohaghegh, *Shale Analytics*, DOI 10.1007/978-3-319-48753-3_7

of the limitations that have to do with specific flow regimes or nuances associated with operational inconsistencies have been tolerated and sometimes clever methods have been devised to get around them. However, when it comes to analysis of production data from shale, a new set of characteristics that may not have been as dominant in the past (in more conventional wells), stand out and undermine the use of any and all production data analysis techniques that rely on traditional statistics-based curve fitting.

What is new and different about production from shale is the impact and the importance of completion practices. There should be no doubts in anyone's mind that a combination of long laterals with massive hydraulic fractures is the main driver that has made economic production from shale a reality. So much so that the professionals in the field have started questioning the impact and influence of reservoir characteristics and rock quality in the productivity of shale wells.[1] While the importance of reservoir characteristics is shared between conventional and unconventional wells, design parameters associated with the completion practices in wells producing from shale are the new and important sets of variables that create the distinction with wells in conventional resources. In other words, completion design parameters introduce a new set of complexity to production behavior that cannot be readily dismissed. However, the impact of the completion design parameters is being dismissed (or at best, assumed to be the same across multiple wells), anytime DCA is used to analyze and forecast production from shale wells.

Therefore, there is a new set of inherent and implicit assumptions that are associated with production data analyses in shale when methods such as DCA are used. In previous cases (nonshale) production is only a function of reservoir characteristics while the human involvements in the form of operational constraints and completion design parameters played minimal roles. In such cases, when DCA is applied, the assumption is that differences in wells production and ultimate recovery that is being estimated by DCA are indications of the impact of the reservoir characteristics. As such, DCA is treated as a technique that without elaborating on the reservoir characteristics (since it does not directly takes them into account) uses the DCA coefficients (such as "D" and "b") as a proxy for these characteristics.

In shale, completion design practices play a vital role. By performing traditional statistics-based production data analysis, we are assuming that reasonable or optimum or may be even consistent completion practices are common in all the wells in a given asset. Author's years of experience related to analysis of oil and gas production from shale wells that include thousands of wells from shale plays such as Marcellus, Huron, Bakken, Eagle Ford, Utica, and Niobrara points to the fact that making such assumptions are nothing short of fantasy and has no basis, what so ever, in reality. Production and completion data that have been thoroughly examined in multiple shale plays clearly point to the fact that such assumptions may prove to be quite costly for the operators.

---

[1]Something that used to be conventional wisdom and reservoir engineering common sense is now being questioned, and for good reasons. To understand the impact and influence of reservoir characteristics and how it is impacted by completion practice in shale, please see Chap. 5 of this book.

However, given the fact that DCA is still used quite widely in the industry, authors of this chapter have set to use Shale Analytics in order to shed some light on the use of DCA and the impact of different reservoir and completion characteristics on the Estimated Ultimate Recovery values that are calculated using different DCA techniques. But first, let us examine different technologies that have been developed recently to make DCA applicable to shale wells. In the following subsections, five most commonly used DCA techniques for shale wells are examined. These techniques are Arp's Hyperbolic Decline (HB), Power Law Exponential Decline (PLE), Stretched Exponential Decline (SEPD), Doung's Decline, and Tail-end Exponential Decline (TED).

7.1.1   Power Law Exponential Decline

Ilk and Blasingame [18] developed the Power Law Exponential (PLE) decline method, which is an improvement of Arps' exponential decline, to get better fit and forecast of EUR specifically for low permeability (unconventional reservoirs). They declared that this technique (PLE) is flexible enough to model transition and boundary-dominated flow by adding the D_∞ term in these equations which handles the late time behavior, often exhibiting boundary-dominated flow. Following equations summarize the Power Law Exponential technique.

Loss Ratio in Power Law Exponential Technique.

$$D = D_\infty + nD_i t^{n-1} \tag{7.1}$$

Power Law Exponential derivative of loss ratio.

$$b = \frac{d}{dt}\left(\frac{1}{D}\right) = \frac{-n(n-1)D_i t^{n-1}}{D^2} \tag{7.2}$$

Power Law Exponential rate–time relation.

$$q = q_i \exp\left(-D_\infty t - \frac{D_1}{n} t^n\right), \tag{7.3}$$

where
   q_i = rate intercept at $t = 0$
   D_∞ = decline constant at infinite time
   D = decline constant
   n = time exponent

7.1.2 Stretched Exponential Decline

Stretched Exponential Decline was introduced by Valko [19]. He developed a new equation which has a different technique to handle the exponent value. Valko stated that the two most significant advantages of this method are the bounded nature of EUR from any individual wells and the straight line behavior of the recovery potential expression versus the cumulative production.

Another aspect announced by Valko is the favorable mathematical properties of this model to deal with large amount of data without subjective deletion, interaction, or data modification, which are important in data-intensive analysis. Based on Valko, all these characteristics give the SEPD method a more rational estimate of EUR compared with other methods.

Stretched Exponential Decline rate–time relation.

$$q = q_i \exp\left[-\left(\frac{t}{\tau}\right)^n\right] \tag{7.4}$$

7.1.3 Doung's Decline

A new DCA was introduced by Doung [21] in 2010. He mentioned that the conventional DCA such as Arps' method overestimates the EUR for ultralow permeable reservoirs or shale wells. Fracture dominated flow regime is more frequently seen in hydraulically fractured shale wells however boundary-dominated flow (BDF) regime will not occur until after many years of production. The drainage area and matrix permeability do not form in the absence of the BDF and pseudoradial. It demonstrates that the fracture network has a dominant contribution to the flow regime, compared to the matrix permeability contribution. Therefore, the conventional models of drainage area are unable to accurately estimate the expected EUR.

Duong introduced his approach for the determination of EUR from wells with dominant fracture flow and insignificant matrix contribution. The local stress changes under fracture depletion would reactivate faults and discontinuities in the rocks, resulting in an increase in the density of connected fractures over time. This expansion in connected fracture network significantly supports the fracture flows. Reactivation of these existing discontinuities will ease enhanced fluid mitigation by increasing the permeability of the rocks.

In a situation with a constant flowing bottom-hole pressure, regardless of the type of fractures and faults, Doung observed a linear relation with unity slope for the log–log plot of cumulative production rate versus time. However, in actual field operation, slopes greater than one is more dominant. If the production data do not follow the straight line, the well reaches its boundary-dominated flow. Doung assumes that the production trend will not curve from the straight line and shale wells have infinite linear flow.

As mentioned before this concept can be supported by fracture expansion over time. Therefore, Duong established a relationship between cumulative production and time, based on the initial gas rate as well as the slope and intercept of this straight line relation. The slope of the line corresponds to "$-m$", and the intercept to "a". The equation is defined as

Doung's relationship between cumulative production and time.

$$\frac{q}{G_p} = at^{-m} \qquad (7.5)$$

The production history of a horizontal Marcellus well is presented in semilog plot in Fig. 7.1 that shows an example of the log–log plot of q/Q_p to obtain "a" and "m" using a linear regression analysis for a Marcellus shale well. The coefficient of variation (R^2) is used to show how close the data is to the fitted line, and it is recommended to accept R^2 values over 0.90.

The next step is the determination of q_1 by plotting gas rate against $t(a, m)$ using Eq. (7.6). A straight line can fit the data and provides us with a slope of q_1 and intercept of q_∞ (rate at infinite time) which can be zero, positive, or negative. It has been discussed by Duong that the original model of q versus $t(a, m)$ plot has no intercept but it cannot be correct for all cases due to operating conditions. Duong's production rate can be calculated by the following equation:

Doung's relationship for calculating production rate.

$$q = q_1 t(a, m) + q_\infty, \qquad (7.6)$$

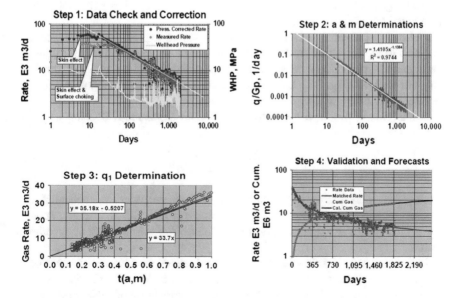

Fig. 7.1 Determination of "a" and "m" Doung's method

where
Time axis in Doung's Equation.

$$t(a, m) = t^{-m} e^{\frac{a}{1-m}\left(t^{1-m}-1\right)} \tag{7.7}$$

Values of q_1 and intercept of q_∞ can be estimated from the graph (Step 3 in Fig. 7.1) in using linear regression for a horizontal Marcellus well. Once q_1 and intercept of q_∞ are determined, gas production forecast can be performed using Eq. (7.6). For the cases that q_∞ equals to zero, estimated ultimate recovery can be calculated from Eq. (7.8):

Estimated Ultimate Recovery using Doung's technique.

$$Q_p = \frac{q_1 t(a, m)}{at^{-m}}. \tag{7.8}$$

7.1.4 Tail-End Exponential Decline (TED)

Arps' Hyperbolic decline curve is widely used for reserve calculation in conventional wells. However, this method results in unrealistically high values for reserves when applied to tight formations including shale. To tackle this issue, different alternative decline curve methods have been proposed. Robertson [67] suggested tail-end exponential decline which provides more conservative forecasts than those by Arp's decline. Figure 7.2 demonstrates the comparison of cumulative production estimates generated by exponential, hyperbolic, and tail-end exponential decline techniques for a typical well.

Tail-end exponential decline is a combination of exponential and hyperbolic decline curves. The curve starts with decline rate of D and decreases over time. At some point hyperbolic curve switches to exponential decline, to show the later stages of well life (Fig. 7.3). In tail-end exponential decline, production rate (q) can be calculated using Eq. (7.9).

Calculation of production rate for the Tail-End Exponential Decline.

$$q = q_i \frac{(1 - \beta)^b \exp(-Dt)}{(1 - \beta \exp(-Dt))^b}, \tag{7.9}$$

where β is responsible for transition from hyperbolic behavior to exponential. For exponential and hyperbolic curves β is in the range of $0 \leq \beta \leq 1$. The parameter "D" is the asymptotic exponential decline rate and "b" is the exponent of hyperbolic decline.

The deficiency of this method can be the assumption value of asymptotic exponential decline rate which need to be determined in advance. This parameter is often chosen by reservoir engineers based on their experience on different wells and

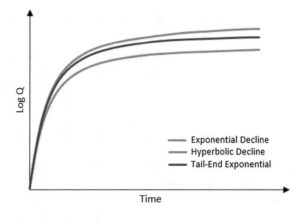

Fig. 7.2 Comparison of cumulative production estimated using exponential, hyperbolic, and tail-end exponential declines

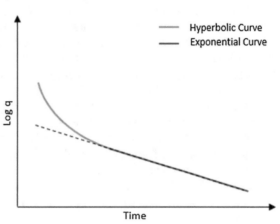

Fig. 7.3 Tail-end exponential decline

reservoirs. After setting a value for the parameter, "D", it is required to obtain values for initial production rate (q_i), "b", and β by fitting a curve on historic production data. The initial production rate can be set as early production rate in actual data. It is assumed that a well will flow in a hyperbolic manner at its early stages of life and exponential at the tail, therefore, other parameters should change in a way that the curve fits the actual data.

After finding a fitted curve for the production data, cumulative production can be calculated by Eq. (7.10):

Calculation of cumulative production for the Tail-End Exponential Decline.

$$Q = q_i \frac{(1-\beta)}{\beta D} \left[1 - \frac{1}{(1 - \beta \exp(-Dt))^{b-1}} \right] \qquad (7.10)$$

7.2 Comparing Different DCA Techniques

The most commonly used technique for reserve estimation and estimating ultimate recovery of a well or a group of wells is DCA. This method is an empirical and graphical method which is used to match historical well production data and extends the curve to forecast the production rate and EUR. In this section, different decline curve methods that were introduced in the previous sections are used to forecast well production and are compared in terms of the range of estimated ultimate recovery for horizontal shale wells.

Since the length of production history of the shale wells may have an impact on the trend of the production forecast and the EUR when different decline curve methods are used, three different lengths of production history (short = 3 months, medium = 12 months, long = 60 months) are considered for horizontal shale wells. The comparison of various DCA methods is shown in Figs. 7.4, 7.5 and 7.6 for short, medium, and long history of data. In this comparison 50-year EUR values are calculated three times for each techniques where each time the amount of production history used to generate the decline curve is changed. It is a well-known fact that the entire process of DCA regardless of the specific technique that is being used is, to a large extent, subjective. Therefore, different engineers may fit the curves in different fashion and one should not expect much precision between analyses performed by multiple engineers.

The 50-year EUR values based on Power Law Exponential, Stretched Decline, Duong decline, and Tail-End Exponential Decline are compared for different lengths of actual historical production and are displayed in Table 7.1 (the percentages next to 12 months in the table indicates changes from 3 months, and the percentages next to 60 months indicates changes from 12 months). Following observation can be made using this table:

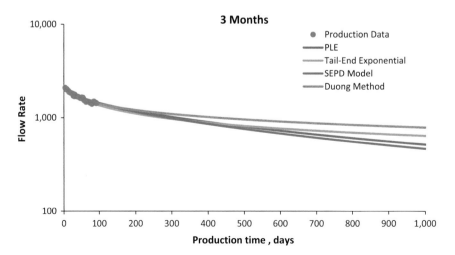

Fig. 7.4 Comparison of production forecast using 3 months of actual production

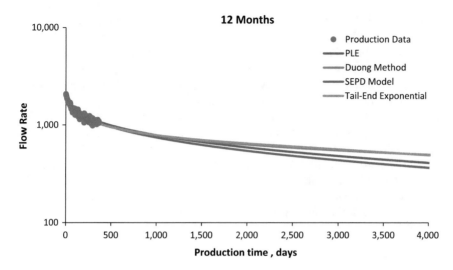

Fig. 7.5 Comparison of production forecast using 12 months of actual production

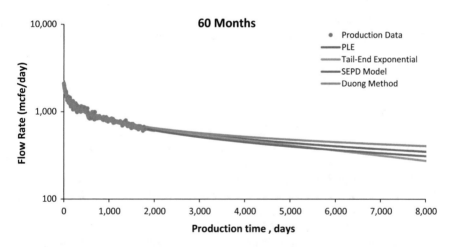

Fig. 7.6 Comparison of production forecast using 60 months of actual production

Table 7.1 EUR comparison for different production lengths using various decline curve methods

50-year EUR (Bcf)								
	TED		SEPD		PLE		Doung	
3 months	3.71		1.68		1.96		7.89	
12 months	5.96	61 %	5.21	210 %	5.89	201 %	7.9	0 %
60 months	5.81	−3 %	6.66	28 %	7.27	23 %	8.24	4 %

1. Duong's method has almost the same EUR values for all three different production lengths and is the one with the most consistency. There is no (or little) difference between 50-year EUR values when 3, 12, and 60 months of actual production history is used to perform the DCA. SEPD and PLE provide the largest amount of discrepancy when the amount of real production history changes from 3 to 12 months and then to 60 months.
2. PLE and SEPD models are not recommended to be used for the cases that short amount of historic production data is available, due to their instability in forecasting.
3. Tail-end exponential remains pretty consistent (and is the most conservative method), once more than 1 year of production history is available. This makes perfect sense since this technique uses hyperbolic decline at the start and exponential decline later in the life of the well.

7.2.1 Is One DCA Technique Better Than the Other?

The fact that there are multiple DCA techniques, brings about a natural question: "Is one DCA technique better than the other?" In order to shed some light on this question and provide some insight on which technique to use under different circumstances we performed some comparative studies. We selected four wells from Marcellus shale with different amount of production history (as short as 292 days and as long as 855 days). We first used the five different techniques to fit the production history.

Figure 7.7 shows the use of five different DCA techniques that are applied to four wells from the Marcellus shale each with a different amount of production history. The four wells shown in this figure include well #10108-1 with 292 days of production history (the top plot), well #10120-1 with 481 days of production history (the second plot from the top), well #10107-4 with 517 days of production history (the third plot from the top), and finally well #10082-5 with 855 days of production history (the bottom plot). In this figure, it is demonstrated that each of the five techniques will fit the production history in a satisfactory fashion. However, the difference between these techniques lies in the way they use the production history in order to forecast 10, 30, and 50-year EUR values.

Results of these analyses are shown in Figs. 7.8, 7.9, 7.10 and 7.11. The general trends in these figures are that the Doung's technique always provides the largest amount of 10, 30, and 50-year EUR (even higher than the normal Arp's hyperbolic technique) regardless of the production history available for analysis and that the TED's technique consistently provides the smallest, most conservative amount of 10, 30, and 50-year EUR. So what is the reason for these discrepancies associated with the EUR forecast of different DCA techniques. Of course authors of these techniques each provide their reasoning on why they have chosen to develop a new technique for DCA of shale wells, but when they are compared to one another, one

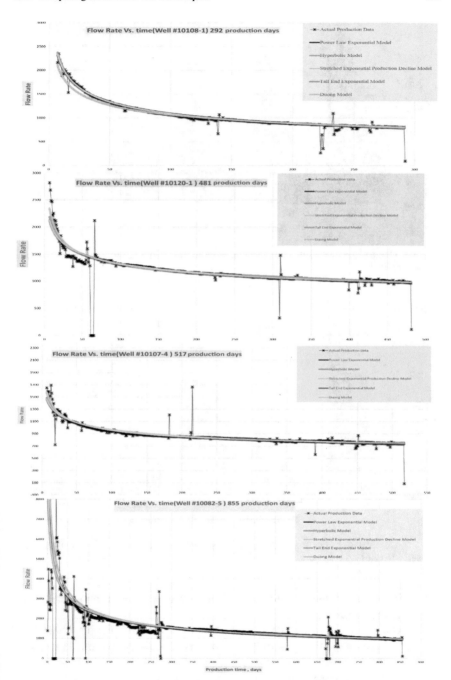

Fig. 7.7 Performing five different DCA techniques on four different wells from Marcellus shale

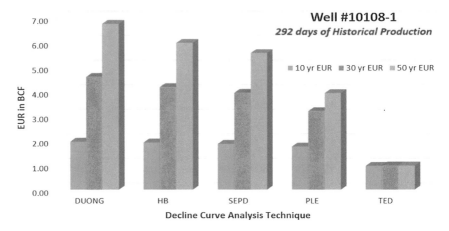

Fig. 7.8 Comparison of the five DCA techniques when forecasting the 10, 30, and 50-year EUR for well #10108-1

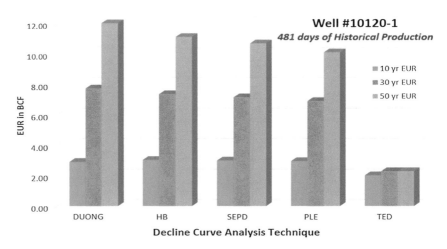

Fig. 7.9 Comparison of the five DCA techniques when forecasting the 10, 30, and 50-year EUR for well #10120-1

can notice the consistency (or lack thereof) through which these techniques estimate the ultimate recoveries.

Our analyses point to the fact that the Doung's technique is the most liberal and the TED technique is the most conservative way of estimating ultimate recovery. This obviously has some mathematical explanation that has to do with the way these techniques choose to treat the coefficients that are used in the form of their equations. In this section we try to translate this into the impact of the reservoir characteristics versus the impact of the design parameters in estimating the ultimate recovery. In other words we try to provide some physical and geological insight

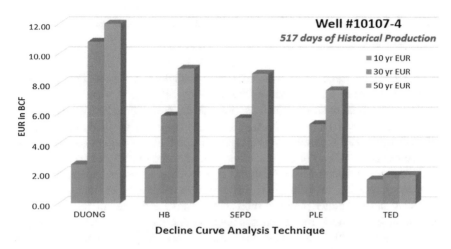

Fig. 7.10 Comparison of the five DCA techniques when forecasting the 10, 30, and 50-year EUR for well #10107-4

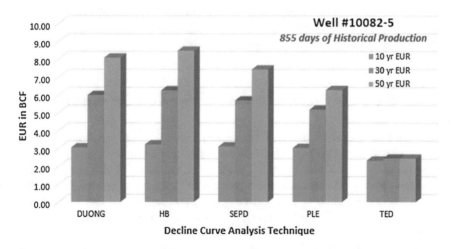

Fig. 7.11 Comparison of the five DCA techniques when forecasting the 10, 30, and 50-year EUR for well #10082-5

into the estimates that are made by each of these techniques and through Shale Analytics identify and demonstrate the impact of different sets of characteristics and the role they play in the degree of conservatism in the estimation of the ultimate recovery from the shale wells.

In other words, we are trying to demonstrate that by using one DCA technique versus another, the analyst (unknowingly, at least up to now) is favoring one set of parameters over another. We are attempting to add some objectivity to DCA that is by nature a subjective technique for the analysis of shale wells.

7.3 Extending the Utility of Decline Curve Analysis in Shale

Since horizontal shale wells that have been completed with multiple stages of hydraulic fracturing are reasonably new, no one really knows which one of these decline curve techniques are actually applicable and are better to be used, when they are estimating the 10–50-year EUR. The major limitation of DCA, including all different methods discussed in this chapter, is that they are neither conditioned to reservoir characteristics nor to completion practices and operational constraints. So the question is whether Shale Analytics can assist us in better understanding the differences between the estimations made by these different techniques.

When one technique provides more conservative estimates of the future production or EUR, does it mean that it is favoring one set of parameters over others? If such questions can be answered, then users of these different techniques may be able to justify why they have chosen to use one DCA technique and not another. Furthermore, using Shale Analytics introduced in this book, the engineer can use actual historical data in order to determine, the degree of which one set of characteristics are dominating others and use such numbers to decide and justify (for the specific field being analyzed) which DCA technique would be more appropriate. In this manner, Shale Analytics can extend the utility of DCA.

7.3.1 Impact of Different Parameters on DCA Technique

Here we first divide all involved parameters into two groups called native and design parameters. Design parameters are those that have been selected by the completion engineers and changing them is in the hand of the engineers and operators. Native parameters are the parameters that are intrinsic to the location where the shale well is drilled and completed and while they can be measured (through well logs and other means) they cannot be modified and are native to the reservoir.

The design parameters include the completion and stimulation design characteristics such as the shot density, perforated/stimulated lateral length, number of stages, distance between stages, number of clusters per stage, amount of injected clean water, rate of injection, injection pressure, amount of injected slurry, proppant amount, etc. The Native parameters are well information and the reservoir characteristics such as porosity, pay thickness, net to gross ratio (NTG), initial water saturation, and total organic content (TOC). In addition, the geomechanical properties such as shear modulus, minimum horizontal stress, young's modulus, and Poisson's ratio are also considered as native parameters.

Figure 7.12 shows the list of native and design parameters that were used for the analyses presented in this chapter. To perform this analysis, the set of parameters shown in Fig. 7.12 is compiled for more than 200 Marcellus shale wells. The

Fig. 7.12 List of native and design parameters collected for the wells in this study

Estimated Ultimate Recovery (EUR) was generated for 10, 30, and 50 years using the five decline curve techniques covered in this chapter [PLE (Power Law Decline), SEPD (Stretched Exponential), HB (Hyperbolic), Duong and TED (Tail-end Exponential)] for all the 200 wells.

In the study presented in this chapter, Shale Analytics that includes Well Quality Analysis (WQA), and Fuzzy Pattern Recognition (FPR)[2] have been used to find hidden patterns that might exist in the relationships between 10, 30, and 50-year EUR and the native and design parameters listed in Fig. 7.12. In other words, WQA

---

[2]These techniques are part of Shale Analytics and are explained in other chapters of this book.

and FPR are employed to find the impact of the native and design parameters on 10, 30, and 50-year EUR.

7.3.2 Conventional Statistical Analysis Versus Shale Analytics

Before displaying the results that can be achieved using Shale Analytics, it is important to present the Conventional Statistical Analysis and demonstrate its inability to shed light on the complexity of the data analysis in shale gas reservoirs. Figure 7.13 demonstrates conventional statistical analysis as it is applied to determining the impact of porosity on the 10-year EUR. In this figure average porosity per well is plotted against the 10-year EUR for 200 wells in the Marcellus shale. This is done in multiple scales (Cartesian, Semilog, Log–Log). Furthermore, the histogram of average porosity values is also shown in this figure.

Figure 7.13 clearly demonstrates that conventional statistical methods do not show any trends. These techniques do not add anything to our understanding of the impact of porosity on 10-year EUR. One may use any type of normalization on these

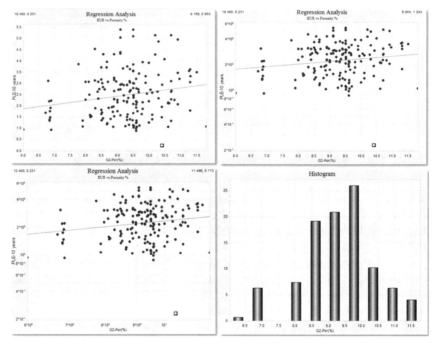

Fig. 7.13 Example of conventional statistical analysis as applied to impact of Porosity on production from shale

plots without any impact on generating or discovering a trend or a pattern. These techniques are incapable of finding any relationship between the average matrix porosity and the 10-year EUR. Should we conclude that such relationships are nonexistent? Most geologists and petroleum engineers know that usually higher porosity results in better production performance and consequently higher EUR. However, such conclusions cannot be deduced from conventional statistical analysis.

When Well Quality Analysis (WQA) and FPR are applied to the same data that were plotted in Fig. 7.13 the results are quite different. WQA, a Shale Analytics Technique, incorporates fuzzy logic to classify wells based on their production charactersitics and then uses the degree of membership in each class and projects them on the parameter being analyzed, in order to detect trends. FPR is the same as WQA, when the number of classes are increased to match the number of wells. Figure 7.14 shows the results of Well Quality Analysis (WQA) performed on average porosity. In this figure it is shown that while the average porosity value for all the wells involved in this analysis is about 9.27 %, the average porosity value for the Poor Wells is 9.05 %, the average porosity value for the Average Wells is 9.42 %, and the average porosity value for the Good Wells is 9.53 %. The trend that is shown in this figure is quite easy to understand and it makes intuitive sense as well. Wells completed in the formation (shale) with higher average porosity values tend to be better producing wells. A well-known geologic and engineering fact can now be confirmed from the actual data collected from the field (field measurements). So, if this technique is capable of confirming something that we already know (although cannot be observed via conventional statistics), maybe we can trust other trends that it can discover, that may not be so intuitive and obvious.

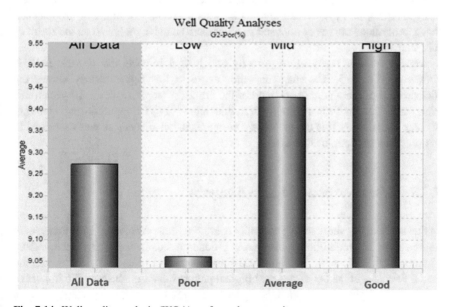

Fig. 7.14 Well quality analysis (WQA) performed on porosity

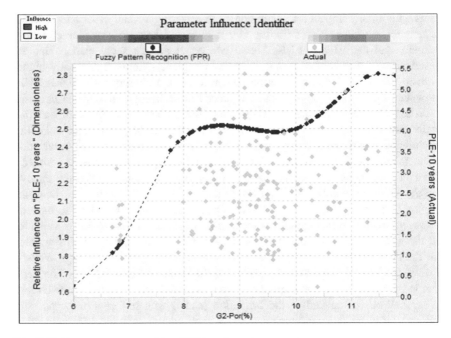

Fig. 7.15 Fuzzy pattern recognition (FPR) performed on porosity

While Well Quality Analysis (WQA) divides the number of wells in the dataset into three (poor, average, and good), four (poor, average, good, and very good), or five (poor, average, good, very good, and excellent) groups (please see Figs. 3.28, 3.29 and 3.30), in FPR we can divide the wells into a large number of groups such that a continuous curve (as opposed to a bar chart) can be generated as the result of this analysis.

Results of FPR analysis on porosity and how it impacts the 10-year EUR is shown in Fig. 7.15. The magenta color curve in this figure clearly shows the increasing (but quite nonlinear) trend of the impact of porosity on the 10-year EUR. Please note that the FPR trend shown in this figure is NOT a regression or a moving average calculation. It is the result of a comprehensive analysis as was described in the previous sections of this book.

7.3.3   More Results of Shale Analytics

In this section, we show multiple results using the Shale Analytics' FPR analysis to demonstrate its capabilities in discovering hidden trends in seemingly chaotic data. Figures 7.16, 7.17, 7.18, 7.19 and 7.20 show a series of plots that include both the actual data as well as the discovered hidden patterns that demonstrate the impact of multiple formation and completion parameters on production indicators from a Marcellus shale asset.

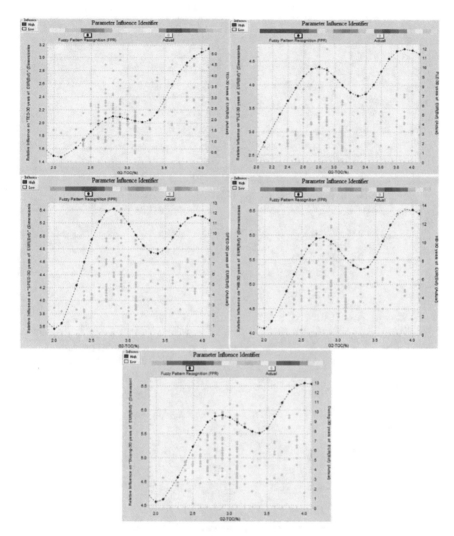

Fig. 7.16 Impact of TOC on 30-year EUR calculated for TED (*top-left*), PLE (*top-right*), SEPD (*middle-left*), HB (*middle-right*), and Doung (*bottom*)

Figure 7.16 includes five plots. Each of the plots in this figure demonstrates the actual TOC (field measurements) versus the 30-year EUR calculated using the five DCA techniques that were covered in this chapter, namely TED (top-left), PLE (top-right), SEPD (middle-left), HB (middle-right), and Doung (bottom). These actual measurements are shown using gray dots (corresponding to the y-axis on the right-hand side of each plot). The magenta color curve in each of these plots shows the impact of TOC on the 30-year EUR as discovered by the Shale Analytics' FPR technology.

In general, the FPR curves in all five techniques show increasing trends that confirm the intuitive notion that wells that have been completed in more mature

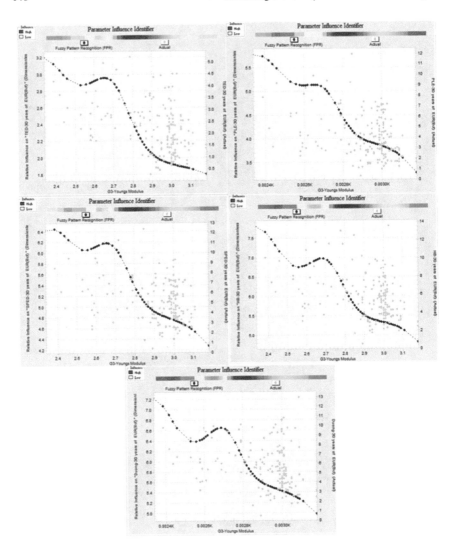

Fig. 7.17 Impact of Young's Modulus on 30-year EUR calculated for TED (*top-left*), PLE (*top-right*), SEPD (*middle-left*), HB (*middle-right*), and Doung (*bottom*)

parts of the reservoir tend to produce more hydrocarbon. However, the trend in each of the plots shows different behavior. The nonlinearity demonstrated in TED's technique (Top-left) is less than other techniques, while the nonlinearity demonstrated in SEPD technique (middle-left) seems to be the most pronounced one, while the rest of the techniques falling in between.

Figure 7.17 is a similar figure showing the impact of a geomechanical characteristic, the Young's Modulus. In this figure, again while all the five techniques show similar (general) behavior, the nonlinearity demonstrated by the five

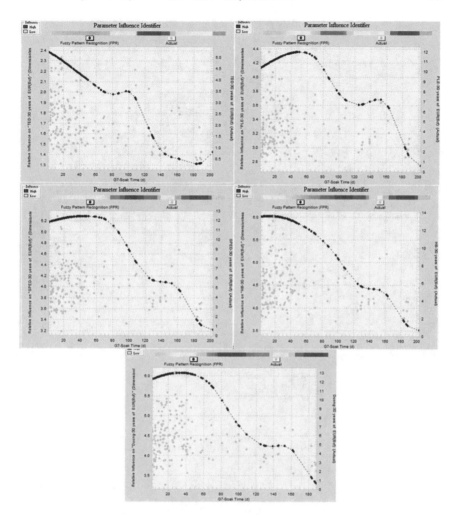

Fig. 7.18 Impact of Soak Time on 30-year EUR calculated for TED (*top-left*), PLE (*top-right*), SEPD (*middle-left*), HB (*middle-right*), and Doung (*bottom*)

techniques are much closer to one another when it is compared to the formation characteristic such as porosity and TOC.

When it comes to completion characteristics such as Soak Time (Fig. 7.18), Number of Stages (Fig. 7.19), and Stimulated lateral Length (Fig. 7.20), the TED technique (top-left) shows more pronounced nonlinear impact when compared to other decline curve techniques. It seems that when it comes to different DCA techniques, if the analyst belongs to the camp that believes formation characteristics have more impact on production than completion characteristics, then he/she should use the TED (most conservative) technique, and if she/he belongs to the camp that believes completion characteristics are the main controlling set of parameters she/he

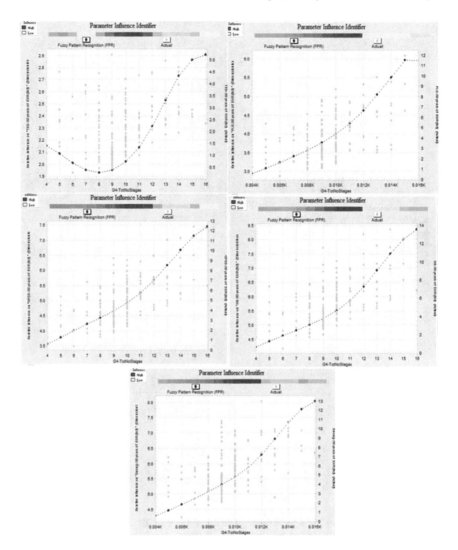

Fig. 7.19 Impact of Total Number of Stages on 30-year EUR calculated for TED (*top-left*), PLE (*top-right*), SEPD (*middle-left*), HB (*middle-right*), and Doung (*bottom*)

should use other decline curve techniques. But what if you do not belong to either of the camps and are interested in finding the truth from your data. We will address this specify issue at the end of this chapter.

Upon completing the advanced data-driven analytics on all involved parameters (well construction, reservoir characteristics, completion, and stimulation), some of which are shown in Figs. 7.14, 7.15, 7.16, 7.17, 7.18, 7.19 and 7.20, Shale Analytics provides much needed insight into how each one of these DCA techniques is impacted by these parameters, albeit implicitly, since none of the authors

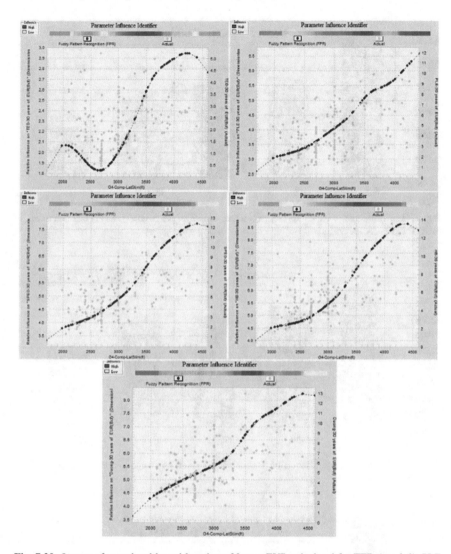

Fig. 7.20 Impact of completed lateral length on 30-year EUR calculated for TED (*top-left*), PLE (*top-right*), SEPD (*middle-left*), HB (*middle-right*), and Doung (*bottom*)

of these techniques have ever included such analyses in their publications. Figures 7.21, 7.22, 7.23 and 7.24 summarize the overall findings of applying Shale Analytics as it applies different flavors of DCA when calculating the shale wells EUR.

Figure 7.21 shows that when TED DCA is used, the analyst (maybe without even realizing it) is putting much more emphasis on the collection of parameters that have been identified in Fig. 7.12 as "Native Parameters". Furthermore, Fig. 7.22 demonstrates that when Doung's DCA is used the analyst (again, without

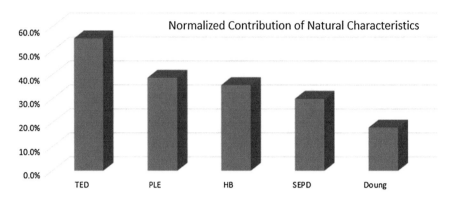

Fig. 7.21 Normalized contribution of parameters classified as natural (native) characteristics to each of the DCA techniques

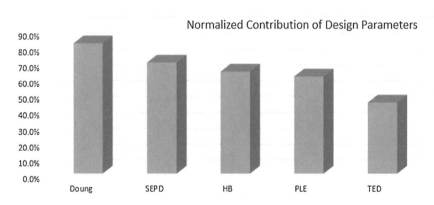

Fig. 7.22 Normalized contribution of parameters classified as design parameters to each of the DCA techniques

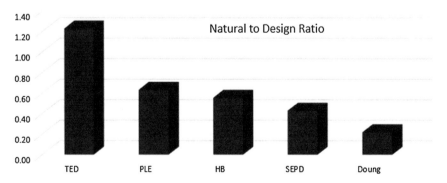

Fig. 7.23 Ratio of the impact of natural characteristics to design parameters for each of the DCA techniques

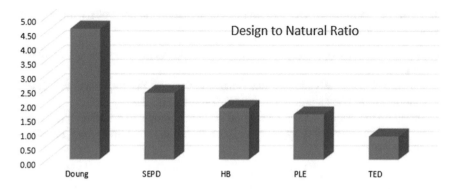

Fig. 7.24 Ratio of the impact of design parameters to natural characteristics for each of the DCA techniques

even realizing it) is putting much more emphasis on the collection of parameters that have been identified in Fig. 7.12 as "Design Parameters".

Figures 7.21 and 7.22 compare the impact of the Native and Design parameters in different DCA techniques, while Figs. 7.23 and 7.24 show the ratio of the impact of the groups of parameters within a given DCA technique. Figures 7.23 and 7.24 quantify the ratio of the impact of the Native Parameters to Design Parameters (Fig. 7.23), and the ratio of the impact of the Design Parameters to Native Parameters (Fig. 7.24) when different Decline Curve Analyses are being used. For instance, in Fig. 7.24, it is shown that when SEPD technique is used, the analyst, without realizing it, is giving twice as much weight to the Design Parameters as compared to the Native Parameters, or when TED is being used during the analysis the Native Parameters are being weighed more than twice as much (in terms of impact on production) as compared to the Design Parameters.

7.4 Shale Analytics and Decline Curve Analysis

In conclusion, it seems that when using techniques such as DCA in order to estimate the ultimate recovery from shale wells, it is not the actual data that is guiding the conclusion of one's analyses, but rather the technique itself. In other words, based on the type of the initial biases and believes (whether you believe Native Parameters are more impactful or Design Parameters are more impactful and each to what degree), you can chose the type of DCA techniques that can support your views. In other words, "bias-in, bias-out".

Based on this observation, and in the opinion of the authors of this chapter, using DCA is a subjective approach that has much more to do with the bias and background of the analyst than with the realities that they are faced with. In order to overcome such shortcomings and let the field measurements guide our analysis (this is the definition of Shale Analytics), the authors recommend the following path to choosing the appropriate DCA technique for horizontal shale wells with massive, multistage hydraulic fractures.

1. Use the detail data from the field and divide them into native and design parameters,
2. Perform Shale Analytics and determine the ratio of the impact of the Design Parameters to the natural parameters (similar to Fig. 7.24),
3. Based on the number that is calculated in Step 2, above, and the bar chart shown in Fig. 7.24, select the appropriate DCA technique to estimate the 10, 30, or 50-year EUR.

Chapter 8
Shale Production Optimization Technology (SPOT)

The goal of this chapter in the book is to demonstrate the utility of Shale Analytics in providing insight into hydraulic fracturing practices in the shale. This is accomplished though application of this technology to a Marcellus shale asset in the northeastern United States. The final results of this technology that is known as Shale Production Optimization Technology (SPOT) include:

(a) Design of optimum multistage frac jobs for newly planned wells,
(b) Estimates of hydrocarbon production for newly planned wells, taking into account uncertainties associated with all reservoir, completion, and stimulation parameters, and
(c) Identification of data-driven, non-biased best completion, and stimulation practices in the Marcellus shale.

The analyses presented in this chapter are performed using Shale Analytics applied to a dataset that include 136 horizontal wells producing from Marcellus shale with more than 1200 stages of hydraulic fractures.

8.1 Dataset

To the best of our knowledge, the dataset used in this study is one of the most comprehensive datasets focused on hydraulic fracturing practices in Marcellus shale that has ever been shared with and consequently analyzed by an independent entity.

8.1.1 Production Data

This dataset includes 136 wells that have been completed in Marcellus shale in the recent years. The dataset includes more than 1200 hydraulic fracturing stages. Some wells have experience up to 17 stages of hydraulic fracturing while others have

© Springer International Publishing AG 2017
S.D. Mohaghegh, *Shale Analytics*, DOI 10.1007/978-3-319-48753-3_8

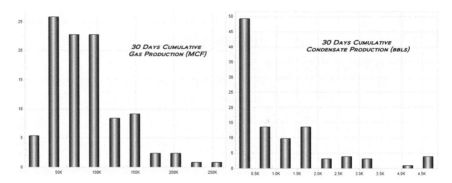

Fig. 8.1 Gas and condensate production; Marcellus Wells

been fractured with as few as four stages. The wells have been producing gas and condensate with an average of 2.8 MMCF/D of gas and 32 bbls/D of condensate at their peak.

As shown in Fig. 8.1, almost three quarters of the wells produce at near average daily gas production (in the first 30 days), while about half of the wells produce condensate that is closer to the lowest level of condensate production in the asset.

The highly productive wells have produced as much as 9.0 MMCF/D of gas and 163 bbls/D of condensate (not the same well) and least productive wells have produced as little as 15 MCF/D of gas and no condensate (not the same well). This shows a significant variation in gas and liquid (condensate) production from these wells. Figure 8.1 shows the distribution of 30 days cumulative gas and condensate production from the wells in the dataset.

8.1.2 Hydraulic Fracturing Data

The diversity of production displayed in Fig. 8.1 is a reflection of the diversity of reservoir characteristics as well as completion and hydraulic fracturing characteristics in these wells. The perforated lateral length in these wells ranges from 1400 to 5600 ft. The wells have been completed and stimulated in up to seven formations with some formations stimulated with up to 17 stages (in the same well) and some formations stimulated with only one stage. The total injected proppant in these wells ranges from a minimum of about 97,000 lbs up to a maximum of about 8,500,000 lbs and total slurry volume of about 40,000 bbls to 181,000 bbls.

8.1.3 Reservoir Characteristics Data

Porosity of one of the existing formations varies from 5 to 10 %, while its gross thickness is measured to be around 43–114 ft with a total organic carbon content

Easting			Average Injection Pressure		
Northing			Average ISIP		
MD			Average Breakdown Pressure		
BTU Area	Group 1	Well Location & Information	Average Maximum Pressure	Group 4	Stimulation Information
Well Type			Average Injection Rate		
Inside Distance (to offset)			Average Maximum Rate		
Outside Distance (to offset)			Average Breakdown Rate		
Percentage Well in Target			Total Fluid Volume		
Average Porosity			Total Slurry Volume		
Net Thickness			Maximum Proppant Contentration		
Average Water Saturation	Group 2	Reservoir Characteristics	Total Proppant		
Average TOC			Average Fracture Gradient		
Average Langmuir Volume			30 Days Cum Gas Production		
Average Langmuir Pressure			30 Days Cum Cond. Production		
Lateral Stimulation Length			90 Days Cum Gas Production		
Shot Density	Group 3	Completion Information	90 Days Cum Cond. Production	Group 5	Production Information
Total Number of Clusters			120 Days Cum Gas Production		
Total Number of Stages			120 Days Cum Cond. Production		
			180 Days Cum Gas Production		
			180 Days Cum Cond. Production		

Fig. 8.2 Data available from the Marcellus shale asset

(TOC) between 0.8 and 1.7 %. The reservoir characteristics of another existing formation include porosity of 8–14 %, gross thickness between 60 and 120 ft and TOC of 2–6 %.

Figure 8.2 shows the list of parameters available in this dataset that includes well location and trajectory, reservoir characteristics, completion, stimulation, as well as production characteristics from 136 completed and producing from Marcellus shale.

8.2   Complexity of Well/Frac Behavior

As it was comprehensively described in the previous chapters in this book, it is a well-known fact that fluid flow in shale is a complex phenomenon that is (at a minimum) characterized by dual-porosity, stress dependent permeability, highly naturally fractured medium that is controlled by both pressure dependent Darcy's law (both laminar and turbulent flow), and concentration dependent Fick's law of diffusion. This is the main reason for the many challenges that are associated with developing numerical reservoir simulation models that can reasonably describe, history match, and forecast production behavior in these formations.

Inclusion of massive, multistage hydraulic fractures only exacerbates the complexity of the modeling process of wells producing form shale formations. As number of producing wells increases in a given field (asset), a phenomenon that is very common in all shale formations throughout the United States, developing full field numerical reservoir simulation models becomes a very complex and time consuming endeavor.

Given all these facts, one should not expect linear and intuitive behavior from such a complex and unconventional reservoir, and the dataset used in this analysis does not disappoint. In this section, we show the complexity of production from this reservoir by displaying the lack of any apparent correlation, pattern of trend between any, and all of the parameters in this dataset.

To demonstrate the aforementioned complexity, all parameters in the dataset were plotted against production indicators such as 30, 90, 120, and 180 days of cumulative gas and condensate production to see if there are any apparent trends or patterns in the data. These plots were made in Cartesian, semi-log, and log–log scales but no trends or patterns were observed. Examples of such plots are shown in Figs. 8.3, 8.4, 8.5, 8.6, 8.7, 8.8 and 8.9 for demonstration.

Although lack of correlation is not surprising, due to the highly complex nature of the production from hydraulically fractured Marcellus shale wells, it may appear quite distressing that this lack of correlation can result in lack of predictability of the production from these wells. This is a legitimate concern since our traditional and conventional approaches to modeling fluid flow in the porous media has not proven to be helpful in these shale formations and has resulted in considerable frustration in many reservoir engineering, reservoir simulation, and modeling and reservoir management circles.

To further examine the possibility of existence of any apparent correlation between different parameters in the dataset with production, pairs of parameters

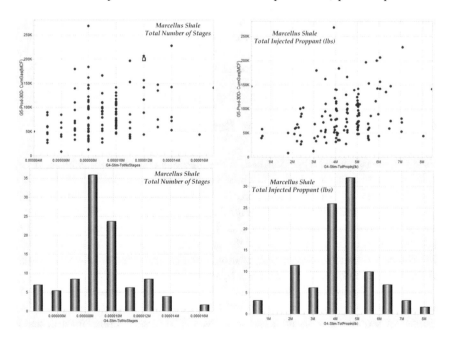

Fig. 8.3 Correlation between total no. of staged (*left*), total inj. proppant (*right*) with the 30 days cumulative gas production

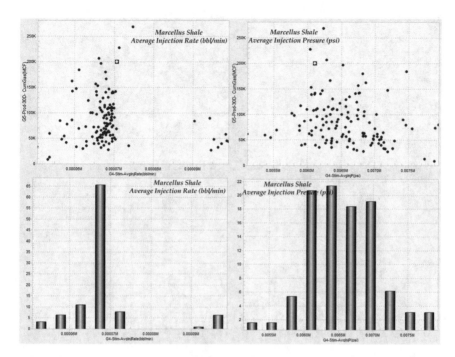

Fig. 8.4 Correlation between inj. rate (*left*), inj. pressure (*right*) with the 30 days cumulative gas production

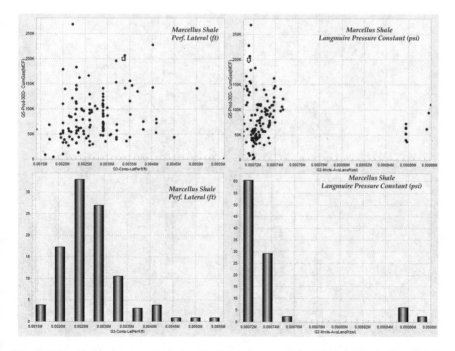

Fig. 8.5 Correlation between pref. lateral (*left*), pressure constant (*right*) with the 30 days cumulative gas production

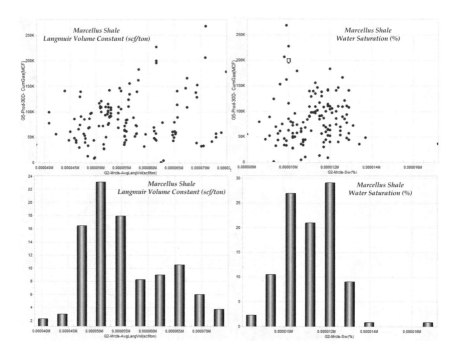

Fig. 8.6 Correlation between volume constant (*left*), water saturation (*right*) with the 30 days cumulative gas production

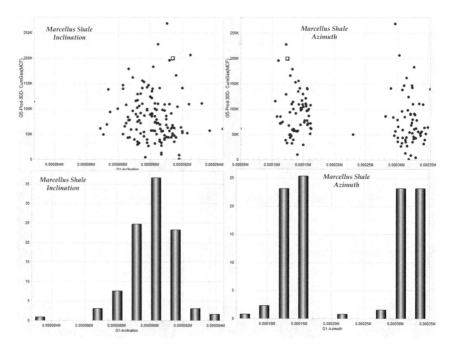

Fig. 8.7 Correlation between inclination (*left*), azimuth (*right*) with the 30 days cumulative gas production

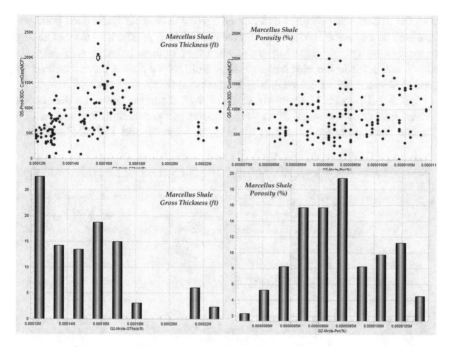

Fig. 8.8 Correlation between gross thickness (*left*), porosity (*right*) with the 30 days cumulative gas production

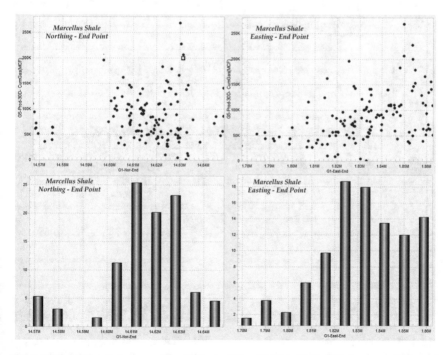

Fig. 8.9 Correlation between easting (*left*), northing (*right*) with the 30 days cumulative gas production

from the dataset were selected and plotted against each other while each point was classified based on the 30 days cumulative gas production. The idea is that although no trend or correlation may be apparent when we plot these parameters against each other (each point in the plot representing a well), if we cluster them based on the general productivity of the well, maybe some correlation will emerge.

Figures 8.10 and 8.11 show six cases of such attempts, while the dataset was divided into three classes of 30 days cumulative gas production. Wells were divided into three groups of less than 50 MM, between 50 and 150 MM and above

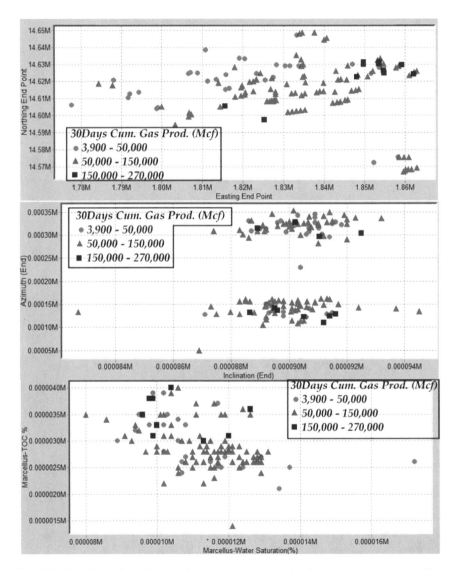

Fig. 8.10 Examining the existence of any apparent correlation between easting and northing (*top*), azimuth and inclination (*middle*) and TOC and water saturation (*bottom*) while classifying wells with different ranges of 30 days cumulative gas production

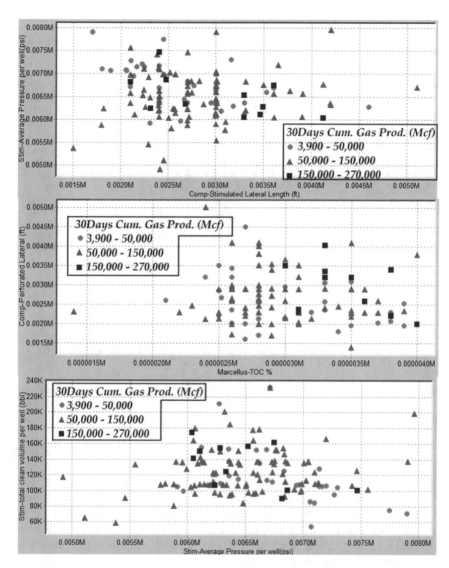

Fig. 8.11 Examining the existence of any apparent correlation between injection pressure and stimulated lateral (*top*), perforated lateral and TOC (*middle*) and injected fluid volume and injection pressure (*bottom*) while classifying wells with different ranges of 30 days cumulative gas production

150 MM of production for the 30 days cumulative gas production. Figures 8.10 and 8.11 show that in all six cases that are representative samples of all possible pairs, poor, average, and good wells are all over the place (there are no trends or patterns in data), no matter which parameters are plotted against each other.

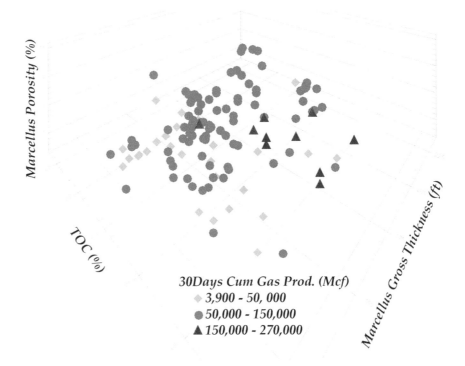

Fig. 8.12 Examining the existence of any apparent correlation between porosity, gross thickness, and TOC of Marcellus Shale while classifying wells with different ranges of 30 days cumulative gas production

Above and beyond the two-dimensional Cartesian, semi-log and log–log plots (with and without grouping the wells based on their productivity), we investigated three-dimensional plotting of the data, again with and without grouping the wells based on their productivity (Figs. 8.12, 8.13, 8.14 and 8.15). This was done since in some cases (tight sands of Rockies) some trends may be observable based on the general categorical productivity of the wells.

Needless to say that none of the efforts to correlate production indicators with well, reservoir, completion, and stimulation characteristics proved to be fruitful. As it is clear from all these figures (Figs. 8.12, 8.13, 8.14 and 8.15), there are no apparent correlations between any of the parameters that are present in the dataset with any of the production indicators that have been calculated for this dataset.

Finally, we used advance statistical techniques such as Analysis Of Variance (ANOVA) to see if we can detect any possible correlation between any of the parameters in the dataset as shown in Fig. 8.16. Not much was concluded from ANOVA. This section demonstrates that the dataset that is representing the hydraulic fracturing practices in the Marcellus shale is so complex that use of conventional statistics in order to understand it and use it to make business decisions is not a fruitful task. Analysis of such a complex phenomenon needs far more

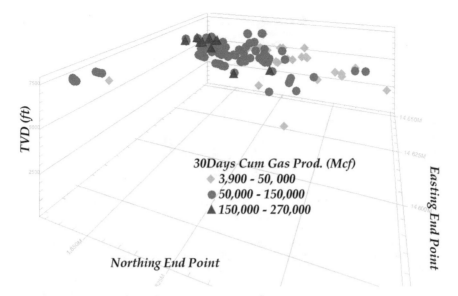

Fig. 8.13 Examining the existence of any apparent correlation between TVD, northing end Point, and easting end point of wells drilled in the Marcellus Shale while classifying wells with different ranges of 30 days cumulative gas production

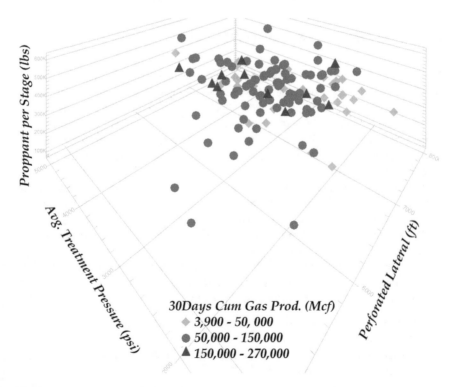

Fig. 8.14 Examining the existence of any apparent correlation between total proppant injected, injection pressure, and perforated lateral of wells drilled in and stimulated in the Marcellus Shale while classifying wells with different ranges of 30 days cumulative gas production

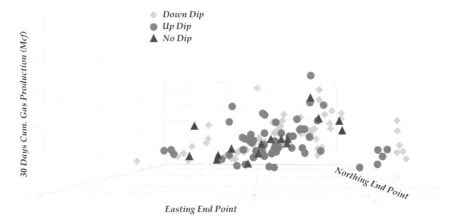

Fig. 8.15 Examining the existence of any apparent correlation between northing end point, easting end point, and the 30 days cumulative gas production of wells drilled in the Marcellus Shale for wells that have been drilled down-dip, up-dip, and with no-dip

sophisticated technologies. We now embark upon application of Shale Analytics to hydraulic fracturing practices in Marcellus shale in order to shed light on this complex phenomenon.

8.3   Well Quality Analysis (WQA)

Well Quality Analysis (WQA) is how we start our Shale Analytics journey with the hydraulic fracturing dataset in shale. We consider Well Quality Analysis to be part of descriptive data mining. This is a process through which the data in the dataset is averaged using the principles of fuzzy set theory [51] and plotted using bar charts in order to reveal some of the hidden patterns in the data. During this process nothing is added or removed from the data. This exercise may be referred to as a unique visualization technique.

Well Quality Analysis (WQA) is a two-step process. In the first step, a target production indicator is selected such that the quality of wells in the field may be defined using this target production indicator. For example in this analysis, we use 30 Days Cumulative Gas Production (30DCGP), to be the target production indicator. Then Fuzzy Set Theory is used to define well qualities based on this production indicator. The range of the 30 Days Cumulative Gas Production in this field has a minimum value of 4MMSCF to a maximum of 270MMSCF. We define four different well qualities within these ranges as follows:

- *Poor Wells:* Poor wells are referred to wells that have 30 Days Cum. Gas between 4MMSCF and 60MMSCF,
- *Average Wells:* Average wells are referred to wells that have 30 Days Cum. Gas between 40MMSCF and 130MMSCF,

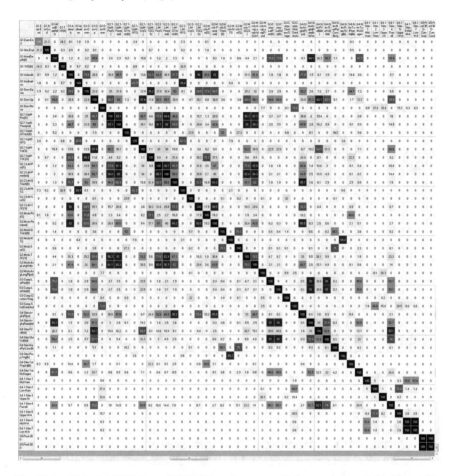

Fig. 8.16 ANalysis Of VAriance (ANOVA) performed on all the parameters in the dataset. The color code helps identify the possible correlations. All parameters are present both in rows and in columns making the main diagonal to present 100 % correlation and shown in the *darkest color*. The *white color* represents 0 correlations and all other colors represent correlation between 0 and 100. The last four parameters (in *rows* and *columns*) are all production indicators, thus showing high correlation. Other neighboring parameters that show high correlations are parameters such as gross thickness and net to gross

- *Good Wells:* Good wells are referred to wells that have 30 Days Cum. Gas between 100MMSCF and 200MMSCF, and finally
- *Excellent Wells:* Excellent wells are referred to wells that have 30 Days Cum. Gas between 180MMSCF and 270MMSCF.

As you might have noticed there are several 30 days production regions that the wells of different qualities overlap. This is one of the unique features of this visualization technology. For example "Poor Wells" and "Average Wells" overlap in the 30 Days Cum Gas Production range of 40MMSCF to 60MMSCF, "Average

Fuzzy Set #	Rise	Top	Top/Fall	Fall
# 1	3986	3986	40000	60000
# 2	40000	60000	100000	130000
# 3	100000	130000	180000	200000
# 4	180000	200000	268288	268288

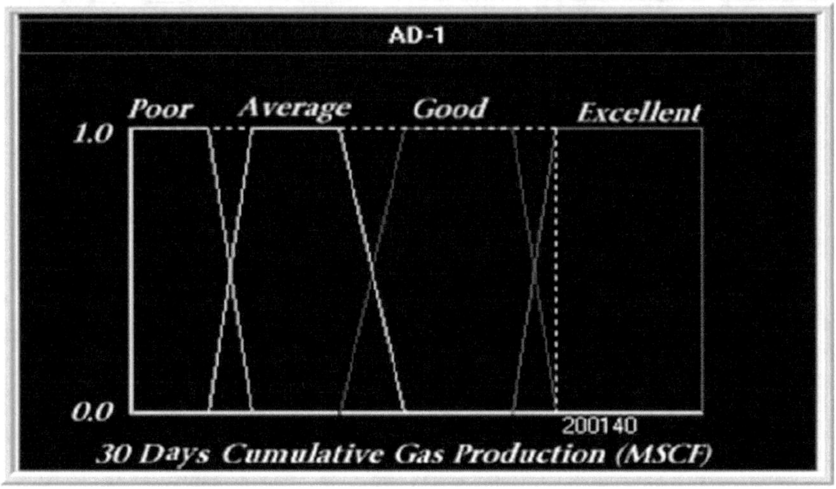

Fig. 8.17 Example of an *"Excellent Well"* based on 30 days cumulative gas production

Wells" and "Good Wells" overlap in the 30DCGP range of 100MMSCF to 130MMSCf, while "Good Wells" and "Excellent Wells" overlap in the 30DCGP range of 180MMSCf to 200MMSCF. These definitions of the quality of wells based on their 30DCGP are shown in Fig. 8.17, while multiple examples for several wells are shown in Fig. 8.18.

Based on this qualitative definition, many wells will have membership in more than one class of wells. For example, while well AD-1 shown in Fig. 8.17 is an "Excellent" well (having a membership of 1.0 in the class of "Excellent Wells") wells that are shown in Fig. 8.18 represent wells of different quality. For example Well #10204 (bottom-right) is both "Good" and "Excellent," each to a degree. This well has a membership of 0.81 in class of "Excellent Wells" and a membership of 0.19 in class of "Good Wells." These classes of memberships are referred to as fuzzy membership functions and are the key to the Well Quality Analysis in descriptive data mining.

Once all wells are defined using this qualitative definition, step two of the process starts. The second step of this analysis, the visualization, is to calculate and plot all the parameters of the dataset based on the fuzzy membership functions calculated in the previous step.

To check the validity of the process we first use this logic and plot the 30 Days Cum Gas Productions. In Fig. 8.19 (plot on the top) existence of a clear trend,

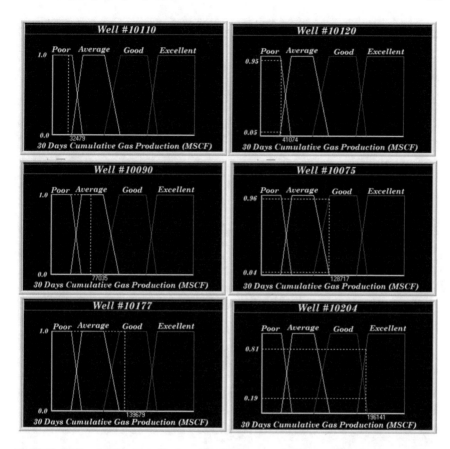

Fig. 8.18 Wells of different quality based on 30 days cumulative gas production

identified by the monotonically increasing red line (from "Poor Wells" to "Excellent Wells"), is quite clear as intended by the classification of the well qualities. It is shown that "Excellent Wells" have the highest 30 Days Cum Gas Production (average 30DCGP of 220 MMSCF) followed by "Good Wells" (average 30DCGP of 140 MMSCF), "Average Wells" (average 30DCGP of 80 MMSCF) and finally "Poor Wells" (average 30DCGP of 38 MMSCF). In this plot, the average 30DCGP of all the wells in the dataset is identified as 85 MMSCF.

The plot at the bottom of Fig. 8.19 shows that 35 % of the wells are classified as "Poor Wells," 72 % are classified as "Average Wells," 28 % are classified as "Good Wells" and finally less than 5 % are classified as "Excellent Wells." Adding all the above-mentioned percentages results in 140 %. These simply mean that at least 40 % of the wells in this field fall somewhere between the identified classes having membership in more than one quality of wells and therefore, include in multiple classes.

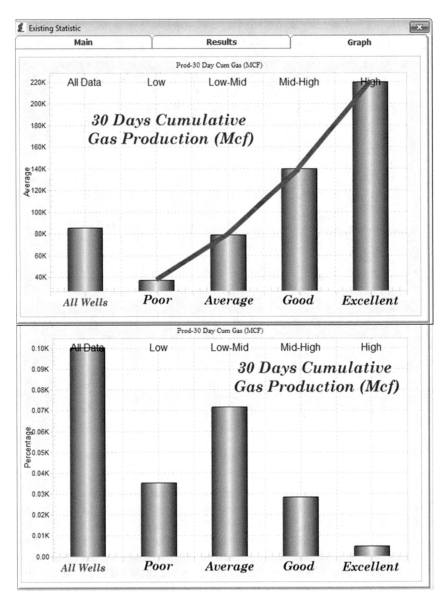

Fig. 8.19 Well quality analysis (WQA) of 30 days cumulative gas production

As it was mentioned earlier, Well Quality Analysis is a process through which the quality of the well (as defined above) is used in order to calculate an average, normalized value for each parameter in the dataset as a function of the wells' fuzzy membership functions in each class of the well categories. The results of this analysis is plotted to see if any specific trends can be observed. As it is

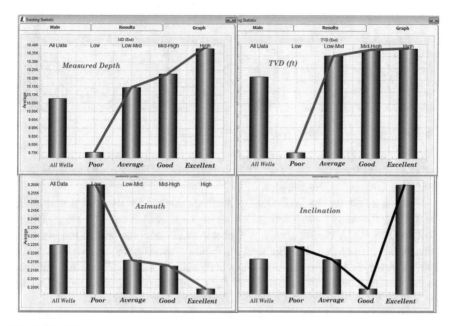

Fig. 8.20 Well quality analysis (WQA) of the well related parameters

demonstrated in this section, from time to time, certain trends (patterns) emerge from this analysis.

Figure 8.20 shows the Well Quality Analysis (WQA) of four parameters related to the well location and trajectory. This plot shows clear trends (identified by red trend line in each plot) in Measured Depth, TVD and Azimuth, while it indicates no apparent trend for (identified by black trend line) Inclination. In the upper left plot it shows that the "Excellent Wells" have an average Measured Depth of 10,375 ft; the "Good Wells" have an average Measured Depth 10,225 ft, while the "Average Wells" have an average Measured Depth 10,150 ft and "Poor Wells" have an average Measured Depth of 9750 ft. This shows a nonlinear but continuously decreasing Measure Depth from "Excellent Wells" to "Poor Wells."

The TVD for the "*Excellent,*" "*Good,*" "*Average,*" and "*Poor,*" wells are 6553, 6552, 6550, and 6100 ft, respectively, and the corresponding values for azimuth are 198, 213, 217, and 260. As shown in the figure the trends are nonlinearly decreasing (from "Excellent" to "Poor") for TVD and nonlinearly increasing (from "Excellent" to "Poor") for Azimuth. The plot on the bottom-right is the WQA for Inclination. Since the trend line for this parameter has an inflection point (it decreases from "Poor" to "Good" and then increases from "Good" to "Excellent"), it means that the correlation of this individual parameter with the 30DCGP is more complex than can be visualized with this technique. To learn more about correlations of such parameters, one needs to study them in the context of Predictive Data Mining, which provides a combinatorial correlation between multiple parameters

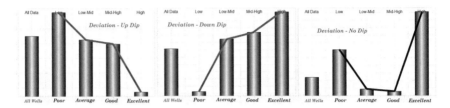

Fig. 8.21 Well quality analysis (WQA) of the Well Trajectory

and the production indicator. Data-Driven Predictive Modeling will be covered in the future sections of this chapter.

Figure 8.21 shows the Well Quality Analysis (WQA) for the well deviation. In the dataset, the wells were identified as having boon drilled Up-Dip, Down-Dip or No-Dip. The WQA clearly shows that in this field early gas production has been unmistakably impacted by the well deviation, favoring the Down-Dip wells. The WQA for the Up-Dip and Down-Dip demonstrate that the "*Excellent*" wells are on the average Down-Dip wells and the "*Poor*" wells are on the average the Up-Dip wells. The WQA for the No-Dip wells displays no clear trends.

Figure 8.22 shows the Well Quality Analysis (WQA) of four completion parameters. This plot shows clear decreasing trends (from "Poor Wells" to "Excellent Wells") in number of clusters per stage and shot density and increasing trends (from "Poor Wells" to "Excellent Wells") in Perforate as well as Stimulated lateral lengths. The upper left plot shows that the "Excellent Wells" have been

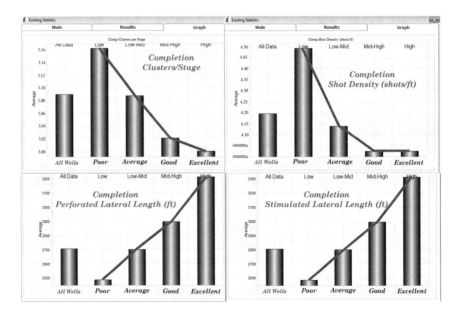

Fig. 8.22 Well quality analysis (WQA) of the completion-related parameters

completed with an average of 3.00 Clusters per Stage; the "Good Wells" have been
completed with an average of 3.02 Clusters per Stage, while the "Average Wells"
have been completed with an average of 3.09 Clusters per Stage, and "Poor Wells"
have been completed with an average of 3.16 Clusters per Stage. This shows a
nonlinear but continuously increasing trend from "Excellent Wells" to "Poor Wells"
for the Clusters per Stage. This trend is repeated for the Shot Density from
"Excellent Wells" to "Poor Wells" ranging from 4.0 to 4.5 Shots per ft.

Figure 8.22 also shows the Well Quality Analysis (WQA) of Perforated and
Stimulated lateral lengths (the two plots at the bottom). The lower left plot shows
that the "Excellent Wells" have an average perforated lateral length of about 3200
ft. The average perforated lateral length decreases for "Good Wells," "Average
Wells," and "Poor Wells," to 2900, 2700, and 2400 ft. Similar trend is observed, as
expected, for the stimulated lateral length.

Figures 8.23, 8.24, 8.25 and 8.26 are dedicated to well quality analysis
(WQA) of the reservoir characteristics of Marcellus shale. In Figs. 5.3 and 5.4
reservoir characteristics of Lower and Upper Marcellus shale are plotted while
Figs. 5.5 and 5.6 show the reservoir characteristics of Marcellus shale as a whole.
For the Lower and Upper Marcellus Shale, six parameters such as Porosity,
Permeability, TOC, Water Saturation, Gross Thickness, and finally "Net to Gross"
are analyzed while for the entire Marcellus Shale Langmuir Volume and Pressure
constants are analyzed as well as the above six parameters.

It is interesting to note that while in the Lower Marcellus shale only two out of
the six parameters (Gross Thickness and NTG) show clear trends, in the Upper
Marcellus shale four out of six parameters (Porosity, Permeability, TOC, and NTG)
show clear trends. The only parameter that is common between these two sets of
reservoir characterizations is the "Net To Gross" ratio. But interestingly enough,
they show opposite trends in each of the formations. In Lower Marcellus, the NTG
"Excellent Wells" have low NTG with an increasing of the NTG toward "Poor
Wells," while in the Upper Marcellus this trend is reversed ("Excellent Wells" have

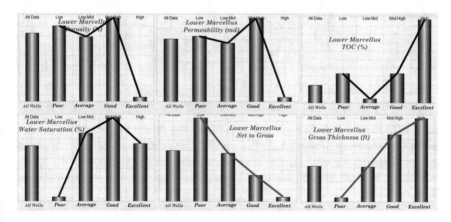

Fig. 8.23 Well quality analysis (WQA) of the lower Marcellus parameters

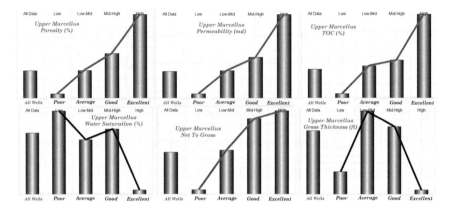

Fig. 8.24 Well quality analysis (WQA) of the upper Marcellus parameters

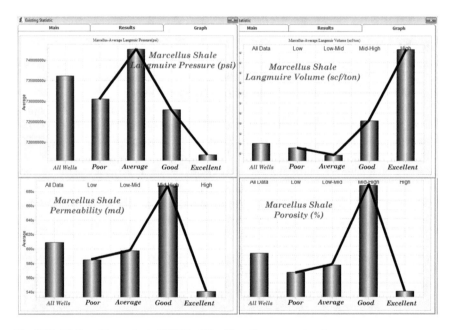

Fig. 8.25 Well quality analysis (WQA) of the Marcellus parameters-1

high NTG with a decreasing trend to NTG of the "Poor Wells"). This analysis points to a necessity to re-evaluate the way NTG is determined. If we have to pick one of these to be correct, the common sense will points toward the Upper Marcellus shale where the quality of the well has a positive correlation with the NTG values.

Furthermore, looking at all the plots in each of the two figures it makes sense to see an increasing trend in well quality as the Porosity and Permeability and TOC of

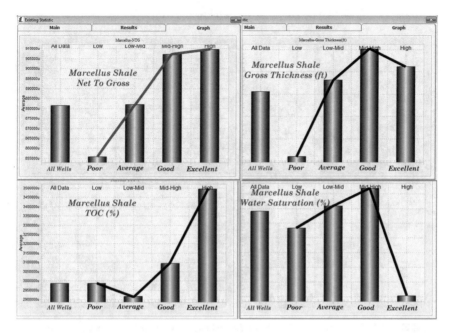

Fig. 8.26 Well quality analysis (WQA) of the Marcellus parameters-2

the formation improves. This trend is detected and observed through the WQA in
the Upper Marcellus and does not show in the Lower Marcellus shale. This fact
further points to the necessity to revisit the values that are furnished in the dataset
for the Lower Marcellus. Another trend that seems to be strange is the behavior of
Gross Thickness for Upper Marcellus shale.

Since Figs. 8.25 and 8.26 (each including four plots) demonstrate the results of
Well Quality Analysis (WQA) performed on the reservoir characteristics of
Marcellus shale as a whole and since the reservoir characteristics of Marcellus shale
as a whole have been calculated by combining the values from Lower and Upper
Marcellus shale, it is expected that the results achieved in these figures to be strange
as well. And the results of Figs. 8.25 and 8.26 do not disappoint. These figures
show that no trend can be detected from any of the reservoir characteristic
parameters.

Figures 8.27 and 8.28 show the results of Well Quality Analysis (WQA) on the
hydraulic fracturing parameters. The eight parameters in these two figures represent
the hydraulic fracturing practices in the Marcellus shale by this operator in the past
several years. Out of the eight hydraulic fracturing parameters presented in these
two figures, five show nonlinear trends that correlate with our definition of well
quality while three do not show any trends. It is interesting that while Total
Proppant in Fig. 8.27 shows a clear trend (better wells are fracked with higher
amount of total Proppant), the amount of Proppant per Stage does not show any
trends (Fig. 8.28). Since the amount of Proppant per Stage is an average value, this

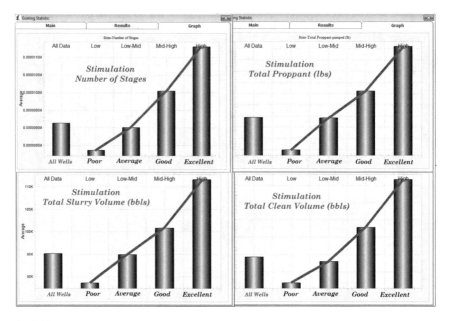

Fig. 8.27 Well quality analysis (WQA) of the hydraulic fracturing parameters-1

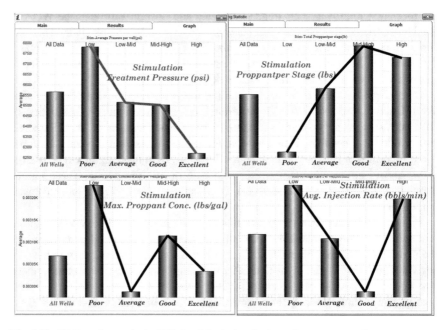

Fig. 8.28 Well quality analysis (WQA) of the hydraulic fracturing parameters-2

points to the fact that similar frac jobs (in a stage) do not necessarily result in the same amount of production.

Given the limitations, in this study we did not consider production on a per stage basis. Since production logging has been performed in this field, performing analysis on a per stage basis is indeed feasible and may shed light on this important issue and allow us to perform frac optimization on a per stage basis. Furthermore, with current use of Distributed Temperature Sensors (DTS) and Distributed Acoustic Sensors (DAS) much more detailed data are available for analysis. DTS and DAS promise to enable and further enrich the analyses that can be provided by Shale Analytics.

In Fig. 8.27 Total Number of Stages, Total Proppant injected, Total Slurry, and Clean fluid volumes show clear trends. Plots in this figure show that increase in each of these parameters indeed correlate with the well quality. In other words, better wells are those that have been fracked with larger number of stages, and consequently have more Proppant and fluid have been used during their completion. Figure 8.28 shows the WQA for injection pressure, injection rate, Proppant concentration, and amount of Proppant per stage. This figure shows that the only parameter (from amongst these four) that shows a clear correlation with well quality is the treatment pressure. The plot (top-left) for treatment pressure in Fig. 8.28 shows that better wells were treated with lower pressure than wells with lower quality. Other three parameters do not show any trend with well quality.

8.4   Fuzzy Pattern Recognition

In Well Quality Analysis, if the total number of well quality categories (classes) are increased such that each well can be placed it its own category (maximizing the granularity), then the Well Quality Analysis (WQA) will turn into Fuzzy Pattern Recognition (FPR). Fuzzy Pattern Recognition (FPR) is an extionsion of Fuzzy Set Theory, and Fuzzy Logic into data analysis in order to deduce understandable trends (continuous patterns) from seemingly chaotic behavior. FPR can be implemented in multidimensions and can take into account impact of multiple parameters in two-dimensional plots.

Figures 8.29, 8.30, 8.31, 8.32, 8.33, 8.34 and 8.35 show the result of application of Fuzzy Pattern Recognition algorithm to all the parameters in the dataset. Figure 8.29 shows the application of FPR to parameters related to the well location and well trajectories such as Northing and Easting end points, Azimuth, and Inclination. Each of the plots in this figure shows the correlation of these parameters with the 30 days cumulative gas production. In this figure, the actual data is also plotted (gray dots corresponding to the y-axis in the right-hand side) in Cartesian scale for reference. The scatter of the actual data demonstrates the lack of any apparent correlation between these parameters with the 30 days cumulative gas production. On the other hand, the FPR displays well-behaved curves that can be explained rather easily.

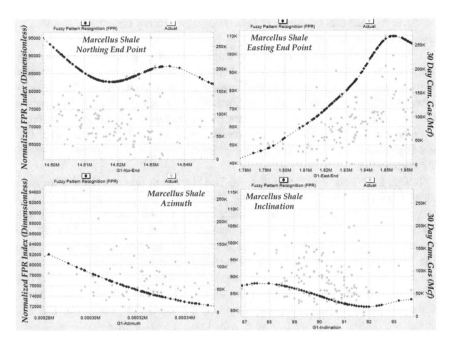

Fig. 8.29 Fuzzy pattern recognition reveals hidden patterns in well locations and trajectories

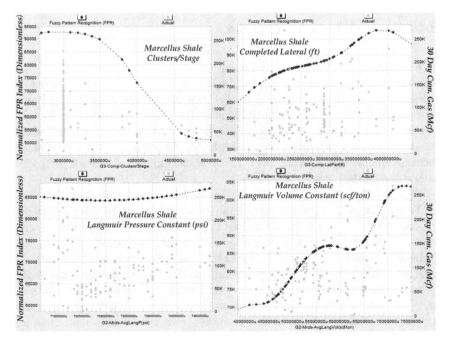

Fig. 8.30 Fuzzy pattern recognition reveals hidden patterns in completion and Langmuir constants

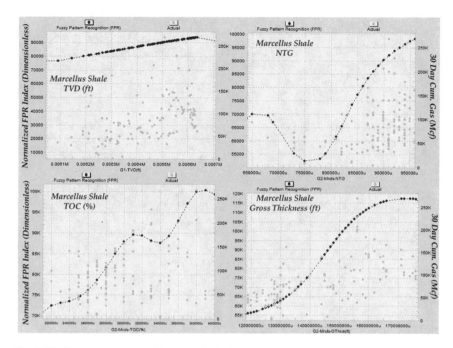

Fig. 8.31 Fuzzy pattern recognition reveals hidden patterns in TVD, TOC, NTG, and Marcellus gross thickness

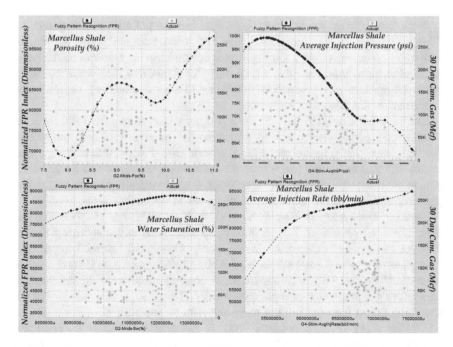

Fig. 8.32 Fuzzy pattern recognition reveals hidden patterns in porosity, initial Sw, injection rate, and injection pressure

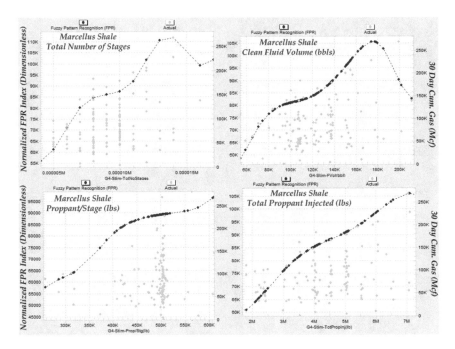

Fig. 8.33 Fuzzy pattern recognition reveals hidden patterns in stimulation parameters such as Total number of stages, clean fluid volume, and proppant per stage, and total injected proppant

For example in the top-left plot where Northing end point is plotted against the 30 days cumulative gas production, the FPR (dark-magenta points forming a well-behaved curve) shows a slightly decreasing and waving behavior as the wells go from north to south. In other words, this plot shows that there seem not to be much changes in gas production behavior (30 days cumulative) as we move southward in this Marcellus shale asset.

On the other hand, the plot on the top-right, which shows the FPR curve of Easting end point, clearly shows a preferential trend toward gas production (30 days cumulative) as we move from west to east. Fuzzy Pattern Recognition curves for Azimuth and Inclination do not necessarily show much of trend. Although, one may observe that the lower azimuth correlates with higher gas production. More on this trend will be discussed in the next sections.

Looking at the plots in Figs. 8.29, 8.30, 8.31 and 8.32, one observes that the FPR curves of several of the parameters do not reveal any convincing trends. Parameters such as "Inclination" in Fig. 8.29, "Completed Lateral Length" and "Langmuir Pressure Constant" in Fig. 8.30, "TVD" in Fig. 8.31 and "Water Saturation" in Fig. 8.32 show minor or no changes in behavior as the 30 days cumulative gas production changes significantly. Even though, it is notable that the Cartesian plot of these parameters leave one with no conclusions.

On the other hand, parameters such as "Easting end point" in Fig. 8.29, "Langmuir Volume Constant" in Fig. 8.30, Marcellus "TOC" and "Gross

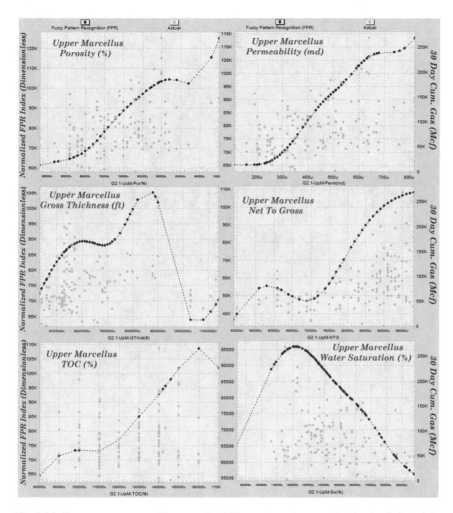

Fig. 8.34 Fuzzy pattern recognition reveals hidden patterns in reservoir characteristics of the upper Marcellus Shale

Thickness" in Fig. 8.31, and finally Marcellus "Porosity" in Fig. 8.32, show significant positive correlations with changes in the 30 days cumulative gas production. In these figures, the only parameters that display a significant inverse correlation with the 30 days cumulative gas production are number of "Clusters per Stage" in Fig. 8.30 and "Injection Pressure" in Fig. 8.32. One also may detect a minor inverse correlation between "Northing End Point" in Fig. 8.29 and the 30 days cumulative gas production.

Figure 8.33 shows the FPR curves for four stimulation parameters. In this figure, the FPR curves for "Proppant per Stage," and "Total Proppant" show a clear and significant (in the case of Total Proppant almost linear) positive correlation with the

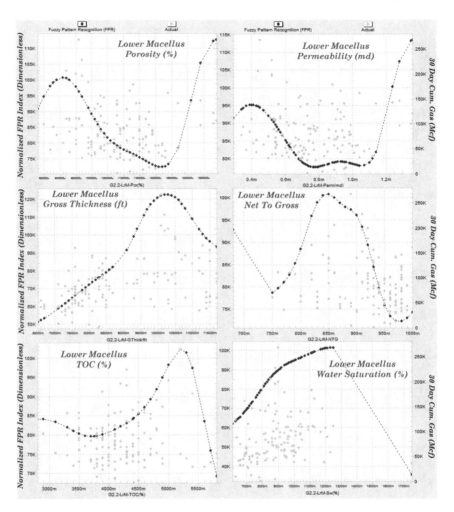

Fig. 8.35 Fuzzy pattern recognition reveals hidden patterns in reservoir characteristics of the upper Marcellus Shale

30 days cumulative gas production while the other two parameters (Total Number of Stages and Clean Fluid Volume) show a slight change in the trend at the end part of the curve. This observed change in the trend in the case of Total Number of Stages and Clean Fluid Volume may be a result of the significant frequency differences in the data and will be addressed in the next sections of this analysis.

Figures 8.34 and 8.35 are dedicated to the reservoir characteristics for each sections of the Marcellus Shale in the dataset, namely Upper and Lower Marcellus. It is notable to see the differences between the FPR curves for the Porosity and Permeability of the Upper and Lower Marcellus Shale in these figures. Of course since permeability has been calculated as a function of Porosity, the dependency of these behaviors on one another is expected. Nevertheless, by concentrating only on Porosity one can observe the major difference in the trends of these curves. While the Porosity (and as a result Permeability) in the Upper Marcellus Shale show a clear positive correlation with the 30 days cumulative gas production, the trend for the Lower Marcellus can be interpreted as either inverse of at least non-existence.

This difference (almost opposite in some cases) in correlation is observable in all reservoir characteristics that are shown in Figs. 8.34 and 8.35 for the Upper and Lower Marcellus. These significant differences may be attributed to contribution of one of these classes of formations to gas while the other may be contributing more to condensate production. More similar analysis should be performed in order to address this specific question. Understanding these behaviors may have important consequences on how the future wells in Western Pennsylvania should be completed and stimulated in order to take full advantage of such behavior in managing this reservoir.

The scale of FRP in all these plots is dimensionless and relative and they have been normalized. The important point about these FPR curves is that they are extracted from seemingly chaotic behavior as shown by the actual data plotted in the same curves (gray dots corresponding to the y-axis on the right-hand side); nevertheless they can display clear trends that can be detected effortlessly.

There are *Two Important Issues* that need to be addressed at this point:

1. These figures (Figs. 8.29, 8.30, 8.31, 8.32, 8.33 and 8.34, 8.35) show the trends of each individual parameter as it correlates with the 30 days cumulative gas production. In other words, the impact of the parameters on one another as they impact the production (a.k.a. combinatorial analysis) is not being observed with FPR curves shown in these figures. These parameters may have either raising or damping impact on the overall trend while their influence on one another is also taken into account. The combinatorial analysis is presented in the later sections of this document.

2. Almost all of the analyses in this document have been performed using the 30 Days Cumulative Gas Production as target (analysis output). Given the fact that:

    (a) Major decline in production from Shale formation usually takes place early in the life of a well due to naturally fractured nature of these formation, and

    (b) These wells produce significant amount condensate along with the gas,

These analyses must be repeated for condensate production and also for different length of both gas and condensate productions, in order to produce meaningful from a reservoir management point of view.

Once these comprehensive analyses (for gas and condensate production and for multiple length of production) have been performed, then the overall results should be combined in order to perform an overall, integrated Data Mining analysis with important reservoir management consequences.

In Figs. 8.36 and 8.37 two examples are presented where trend visualization capabilities of Well Quality Analysis (WQA) are demonstrated. In these figures, the Cartesian scatter plots of the parameters (Perforated Lateral Length in Fig. 8.36 and

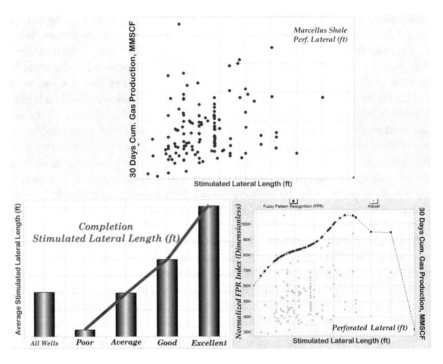

Fig. 8.36 Descriptive data mining for perforated lateral length

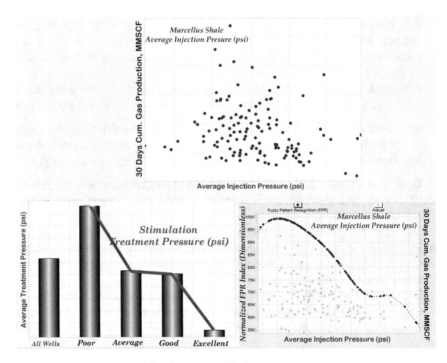

Fig. 8.37 Descriptive data mining for average injection pressure

Treatment Pressure in Fig. 8.37) are shown versus the selected production indicator (30 Days Cumulative Gas Production—30DCGP) on top of each figure. In both cases, the scatter of the data demonstrates no apparent patterns or trends between each of the parameters and the 30DCGP. The plot on the bottom-right shows the Fuzzy Pattern Recognition for each parameter. As mentioned in the previous section, the FPR curves reveal hidden patterns in the data.

The FPR curve in Fig. 8.36 shows a nonlinearly increasing pattern in production behavior as a function of increase in the perforated lateral length. This curve also shows an inflection (change in the slope direction) close to the 4000 ft. Please note that the sharp decrease in production behavior at the end of the graph coincides with a very small number of well (data points) and therefore must be dismissed. The Well Quality Analysis (WQA) that is the bar chart with the red trend line on the bottom-left of the figure shows that the average perforated lateral length for the excellent wells is slightly higher than 3300 ft, followed by 3000 ft for good wells, 2800 ft for average wells and finally a bit higher than 2500 ft for the poor wells. Please note that these values are averages, nevertheless the existence of a monotonically increasing trend is quite clear from the chart.

8.5 Key Performance Indicators (KPIs)

Using the Fuzzy Pattern Recognition (FPR) technology, the contribution (influence) of each of the parameters on any given production indicator (in this case 30 Days Cumulative Gas Production—30DCGP) can be calculated and compared. The result of such analyses is a tornado chart that ranks the impact (influence) of all parameters on the 30DCGP, also known as Key Performance Indicators (KPI).

An example of the result of such analyses is shown in Fig. 8.38. Upon completion of the KPI analyses, the score of the ranking of the parameters can be processed in order to analyze impact of the parameters influence on gas and condensate production as a function the production length. In ranking of the influence of parameters in the KPI analysis, the first (most influential) parameter is given the score of 100 and the score for all other parameters are normalized with respect to the parameter that has ranked the highest. In the example shown in Fig. 8.38, all the parameters in the dataset are analyzed against the 30 Days Cumulative Gas Production. In this analysis, Gross Thickness of Marcellus Shale (parameters identified as Marcellus Shale is the combination of both upper and lower Marcellus in the dataset) is ranked as the most influential parameter followed by the Net to Gross of Upper Marcellus and Gross Thickness of Lower Marcellus. The top five parameters are completed by Easting End Point and Total Number of Stages.

It is interesting to note that of top ten most influential parameters in this analysis six are reservoir characteristics, one is well location and three are stimulation related. Out of the top six reservoir characteristics parameter three belongs to Upper Marcellus (NTG, Porosity and Permeability), two belong to Lower Marcellus

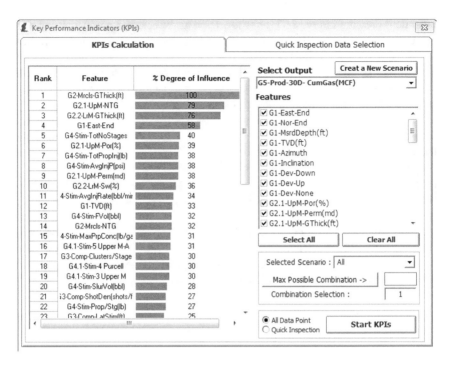

Fig. 8.38 Identifying and ranking the key performance indicators for the 30 Days cumulative gas production

(Gross Thickness and Water Saturation), and one to (integrated) Marcellus Shale (Gross Thickness).

In a separate but similar analysis (Fig. 8.39), reservoir characteristics of Lower and Upper Marcellus were removed and the only set of reservoir characteristics that remained in the analysis were those identified as Marcellus that are an integration of both, Lower and Upper Marcellus. Although this was done in order to give equal weight to the stimulation and reservoir characteristics parameters, it is noteworthy that based on our previous analysis some parameters of Lower and Upper Marcellus display opposite trends when analyzed against 30DCGP (refer to Figs. 8.34 and 8.35). This may result in minimizing the impact of integrated Marcellus reservoir characteristics.

In this new analysis, Marcellus Gross Thickness ranks first (again) followed by Easting end point and Total Number of Stages. Total amount of Proppant injected and Injection Pressure complete the top five parameters. In this case, the top ten parameters is consisted of six stimulation-related parameters (Total Number of Stages, Total amount of Proppant injected, Injection Pressure, Injection Rate, Clean Volume, and Maximum Proppant Concentration), two reservoir characteristics (Gross thickness and Net To Gross), and two well location parameters (Easting End Point and TVD). Figures 8.40 and 8.41 show similar analyses for the 30 Days Cumulative Condensate production.

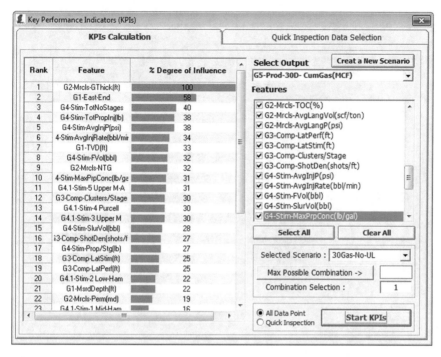

Fig. 8.39 Identifying and ranking the key performance indicators for the 30 days cumulative gas production, removing individually defined lower and upper Marcellus reservoir characteristics

In continuation of these analyses and to demonstrate the capabilities of this technology, the KPI analyses were performed on cumulative gas and condensate productions at 30, 90, 120, and 180 days. Performing these analyses has the potential to identify the impact of each of the involved parameters on gas and condensate production as a function of time. In other words, it would be interesting to see if impact of a certain parameter (or group of parameters such as reservoir characteristics or stimulation-related parameters) is changed as a function of time. Is it possible that some parameters are stronger contributors during the early life of the well and their impact diminished as a function of time as the well enters different phases of its production behavior or visa-versa? Such information may have important ramifications in the economics of production and contribute to reservoir management.

Of course in order for such analyses to be meaningful they should at least be performed during the transient, late transient, and pseudo steady-state periods of a well's life as shown in Fig. 8.42, but to demonstrate the capabilities of such analyses, they we performed, as mentioned before, on cumulative gas and condensate productions at 30, 90, 120, and 180 days.

These analyses identify the *impact of changes in different parameters* (relative to one another), on the gas and condensate productions. In the next several figures that

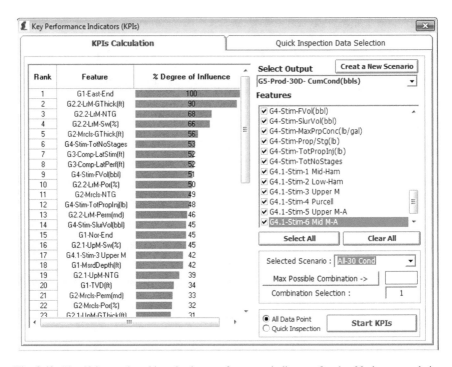

Fig. 8.40 Identifying and ranking the key performance indicators for the 30 days cumulative condensate production

are shown as bar charts, the height of each bar represents the normalized relative influence of each parameter on either gas or condensate production.

Figure 8.43 shows the KPI analyses that are performed for reservoir characteristics. This figure indicates that reservoir characteristics of Lower Marcellus have a more distinct impact on production when compared to those of Upper Marcellus. It must be noted that these types of analyses (relative comparison of influences) mainly point to the influence of changes without identifying the upward or downward direction of the impact. The direction of the impact (increasing or decreasing production) was analyzed in the previous section (Fuzzy Pattern Recognition) and was demonstrated using FPR curves in Figs. 8.34 and 8.35.

It is also evident from Fig. 8.43 that the most influential parameter of all, when it comes to gas production, is change in the gross thickness of lower Marcellus followed by "Net To Gross" ratio and initial water saturation with Total Organic Carbon content playing the least important role when compared with other parameters. This is reasonable since the variability of TOC in our dataset is not large and furthermore, it is expected that the effect of TOC as a contributor to production will start to show later in the life of a well. Furthermore, given the fact that the KPI analysis concentrates on the impact of changes of different parameters on the production, it may be observed that lack of influence of one parameter may

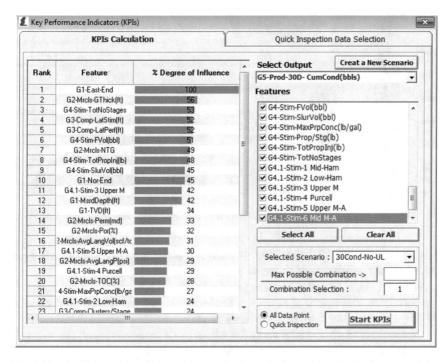

Fig. 8.41 Identifying and ranking the key performance indicators for the 30 days cumulative condensate production, removing individually defined lower and upper Marcellus reservoir characteristics

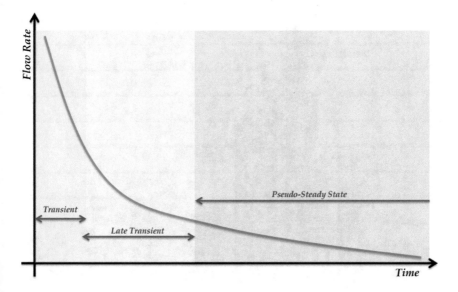

Fig. 8.42 Conceptual flow regimes during the life of a well

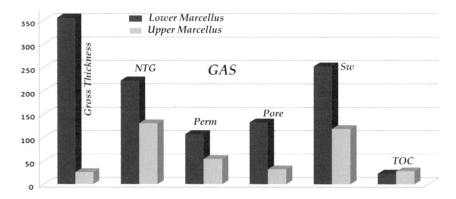

Fig. 8.43 KPI analysis for reservoir parameters for upper and lower Marcellus (gas production)

have to do with the fact that little (or insignificant) amount of change has been recorded for that parameter in the dataset.

Figures 8.44, 8.45 and 8.46 show the results of KPI analysis for all the reservoir characteristics parameters of Marcellus Shale on gas and condensate production at 30, 90, 120, and 180 days of cumulative production.

Figures 8.47, 8.48 and 8.49 show the results of KPI analysis for all the reservoir characteristics parameters of Marcellus Shale on gas and condensate production at 30, 90, 120, and 180 days of cumulative production.

This figure shows that the single most important parameter that has the most impact on the gas production (as far as the well location and trajectory is concerned)

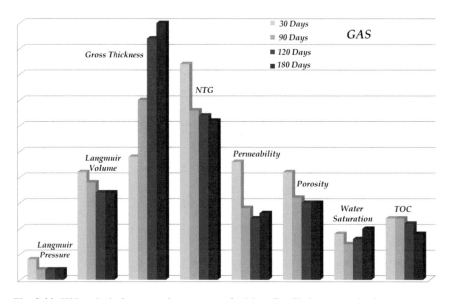

Fig. 8.44 KPI analysis for reservoir parameters for Marcellus Shale (gas production)

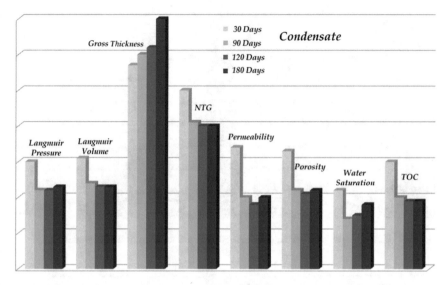

Fig. 8.45 KPI analysis for reservoir parameters for Marcellus Shale (condensate production)

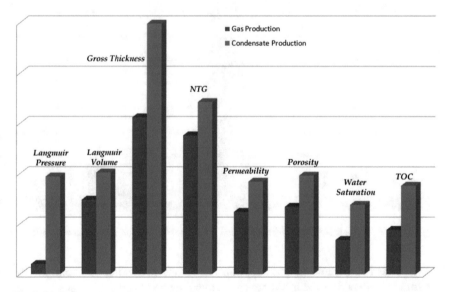

Fig. 8.46 KPI analysis for reservoir parameters for Marcellus Shale (gas and condensate production)

is changes in the Easting end point of the well. Furthermore, it indicates that the importance of the changes of this parameter become more pronounced with the length of time of production (Fig. 8.51).

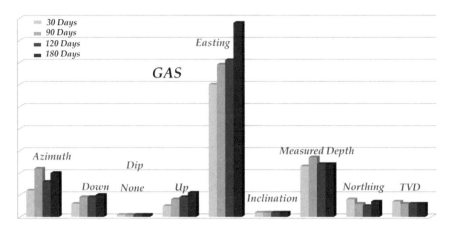

Fig. 8.47 KPI analysis for parameters associated with the well location (gas production)

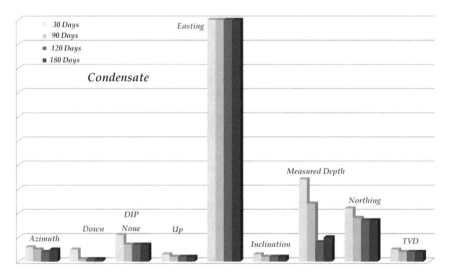

Fig. 8.48 KPI analysis for parameters associated with the well location (condensate production)

The next two parameters that are a distant second to Easting end point are the measured depth and azimuth without any convincing trend about the change of their influence as a function of length of production. It is interesting to note that changes in dipping, inclination, and other parameters play an insignificant role when they are compared to the easting end point. This is an interesting finding that can help in placing the next set of wells in this field. Everything else being equal, the operator should try to move eastward in the field in order to get higher gas production.

Figures 8.50, 8.51 and 8.52 show the results of KPI analysis for all the completion parameters of Marcellus Shale on gas and condensate production at 30, 90,

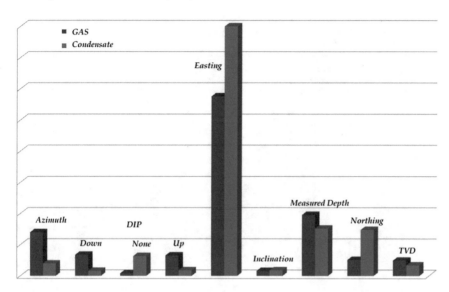

Fig. 8.49 KPI analysis for parameters associated with the well location (gas and condensate production)

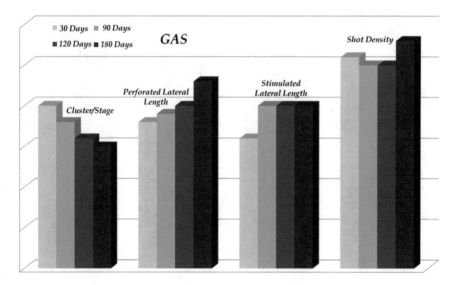

Fig. 8.50 KPI analysis for parameters associated with completion (gas production)

120, and 180 days of cumulative production. Figures 8.53, 8.54 and 8.55 show the results of KPI analysis for all the stimulation parameters of Marcellus Shale on gas and condensate production at 30, 90, 120, and 180 days of cumulative production.

Finally, Figs. 8.56, 8.57, 8.58, 8.59 and 8.60 show the impact of all the parameters, categorized and normalized into groups, on gas and condensate

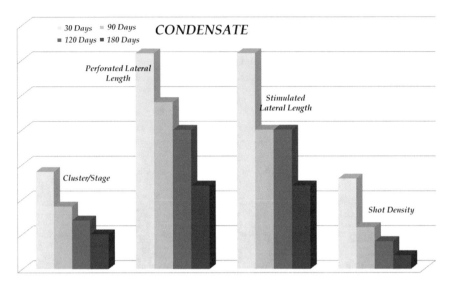

Fig. 8.51 KPI analysis for parameters associated with completion (condensate production)

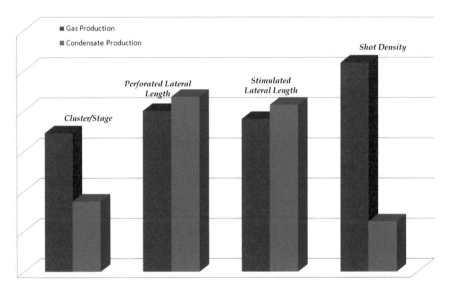

Fig. 8.52 KPI analysis for parameters associated with completion (gas and condensate production)

production. These figures show that in both gas and condensate production changes in lower Marcellus reservoir characteristics have the most impact on production. Another important aspect of these figures is that they show that no individual set of parameters (drilling, reservoir characteristics, completion, and stimulation) or category dominates the influence on production.

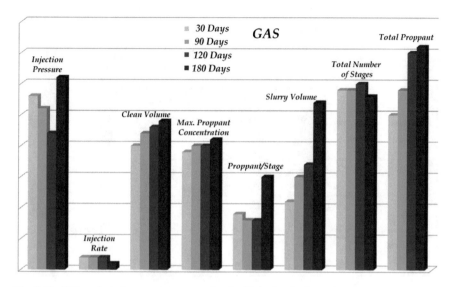

Fig. 8.53 KPI analysis for parameters associated with hydraulic fracturing (gas production)

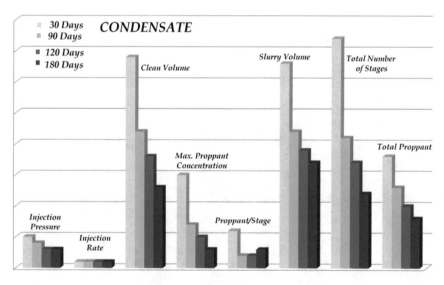

Fig. 8.54 KPI analysis for parameters associated with hydraulic fracturing (condensate production)

Figures 8.59 and 8.60 all the parameters in the dataset have been divided into two groups. Group one are the reservoir characteristic parameters and group two are the completion and stimulation parameters. Reservoir characteristics are the 29 parameters such as Well Location and Trajectory and Marcellus Static parameters.

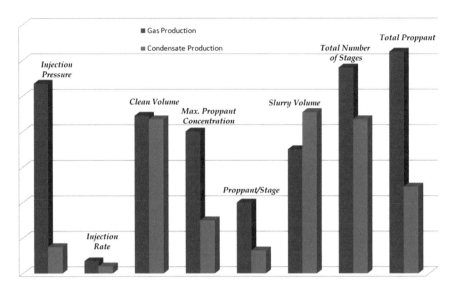

Fig. 8.55 KPI analysis for parameters associated with hydraulic fracturing (gas and condensate production)

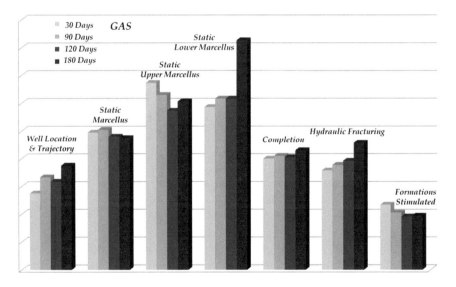

Fig. 8.56 KPI analysis for groups of parameters (gas production)

Completion and Stimulation parameters include the 19 parameters such as Completion and Stimulation parameters.

Figures 8.59 and 8.60 emphasize the importance of reservoir characteristics and where the wells are placed. In other words, sound reservoir management practices

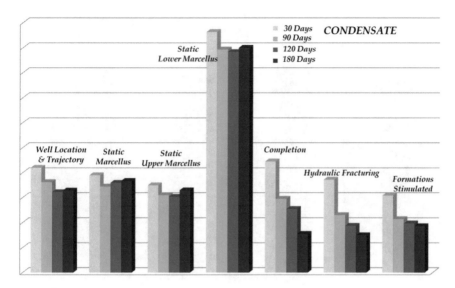

Fig. 8.57 KPI analysis for groups of parameters (condensate production)

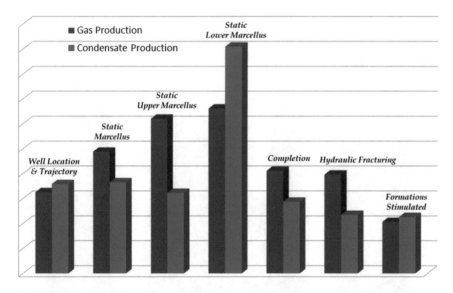

Fig. 8.58 KPI analysis for groups of parameters (gas and condensate production)

can make a Marcellus Shale asset far more attractive than none- engineered and non-optimized production. Sound reservoir management practices can make the economics of drilling in Marcellus shale significantly better.

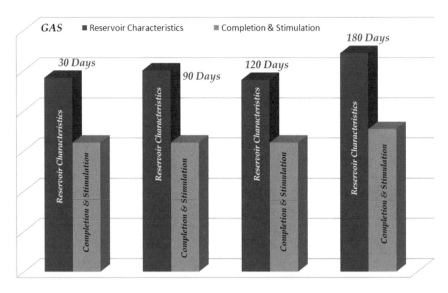

Fig. 8.59 KPI analysis for reservoir versus completion parameters (gas production)

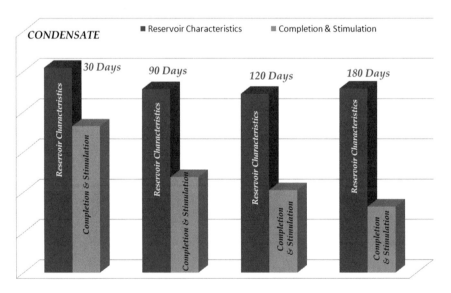

Fig. 8.60 KPI analysis for reservoir versus completion parameters (condensate production)

8.6 Predictive Modeling

The oil and gas industry spent several years grappling with Decline Curve Analysis, Analytical solutions, Rate Transient Analysis, and Numerical Reservoir Simulation, in order to be able to predict production from shale. Complexity of modeling production from shale, a naturally fractured source rock with multiple, coexisting storage and transport characteristics, is exasperated by massive multicluster, multistage hydraulic fractures. Finally, after years of efforts and hundreds of millions of dollars spent, the leaders of the industry have come to the conclusion (and are now openly admitting in panels and conferences) that these conventional technologies have added little value to our understandings and predictive capabilities when it comes to modeling the production from shale. It will take a while before this realization trickle down to other sectors in the industry and become a widely held view.

The question now becomes, "what should we do now, as an industry?" Well, given the fact that the bar for building predictive models have been set pretty low by other technologies, this should not be very hard task to achieve using technologies such as artificial intelligence and data mining, or as we have called them in this book, Shale Analytics. It is only fair to judge the value of the predictive models that are developed using Shale Analytics (data-driven analytics), by comparing it to other technologies that have tried, and overwhelmingly failed, to accomplish the same objective. Therefore, data-driven predictive modeling of production from shale should be viewed in such a landscape, and what is presented in this chapter of the book attempts to satisfy such expectations.

Building a data-driven predictive model of the gas and condensate production from wells that have been drilled, completed and stimulated with massive hydraulic fractures in the Marcellus shale, is the objective of this step of the analysis. It goes without saying that any chance to perform predictive data mining that will eventually result in identification of best stimulation practices in Marcellus shale will depend on our success in developing a verifiable predictive model.

8.6.1 Training, Calibration, and Validation of the Model

To be technically accurate and useful to the operating companies, the predictive model should be able to couple reservoir characteristics with completion and stimulation practices. Figure 8.61 shows the list of parameters that have been used in order to develop the data-driven predictive model in this chapter. A closer look of the list of parameters that are shown in Fig. 8.61 points to several important issues that distinguish data-driven predictive models from other competing technologies.

1. Presence of data that represents many categories other than just production, clearly distinguishes this technology from Decline Curve Analysis that is a

Fig. 8.61 Input parameters
used in the predictive model

Easting End Point
Northing End Point
Marcellus-Porosity%
Marcellus-Permeability (md)
Marcellus-Gross Thickness(ft)
Marcellus-NTG
Marcellus-Water Saturation(%)
Marcellus-TOC %
Marcellus-Average Langmuir Volume (scf/ton)
Marcellus-Average Langmuir Pressure(psi)
Comp-Perforated Lateral (ft)
Comp-Stimulated Lateral Length (ft)
Comp-Clusters per Stage
Comp-Shot Density (shots/ft)
Stim-Average Injection Pressure per well(psi)
Stim-Average Injection Rate Per well(bbl/min)
Stim-total clean volume per well (bbl)
Stim-total Slurry volume per well (bbl)
Stim-Maximum propant Concenteration per
Stim-Total Proppantper stage(lb)
Stim-Total Proppant pumped (lb)
Stim-Number of Stages

curve fit of the production data and adds little value to our understanding of
production characteristics.

2. Inclusion of actual completion and hydraulic fracturing characteristics such as
   shot density, number of clusters per stage, average injection rate and pressure,
   and amounts of fluid and Proppant placed in the fracture, clearly distinguished
   this technology from numerical simulation and Rate Transient Analysis that rely
   on fantasy driven parameters such as fracture half-length, fracture width, height,
   and conductivity.
3. Conditioning the production behavior of wells to this large number of field
   measurements provide the type of predictive model that allows further analysis
   of the model to reveal important information regarding the storage and transport
   phenomena in shale.

As it will be demonstrated, this model can predict 30 days cumulative gas
production as a function of well location, Marcellus shale reservoir characteristics,
well completion and hydraulic fracturing design parameters that were implemented.
Inclusion of these parameters is important since it was clearly demonstrated in the
previous sections of this book that shale reservoir characteristics influence the

30 days cumulative gas production just as much as (and may be even more than) completion and hydraulic fracturing parameters do. In other words, performing the same completion practices (as a cookie cutter approach) in all wells regardless of their location (representing the reservoir characteristics) is not a good and advisable practice. Furthermore, our studies of the data from more than 3000 wells completed in the Marcellus, Utica, Eagle Ford, Niobrara, and Bakken shale have shown that sound reservoir management practices can make a marked difference in economic development of these assets.

To maximize the possibility of developing a robust data-driven predictive model a strategy is employed that maximizes the capabilities of the model while making sure that the known physics of the process (no matter how general and qualitative) is honored, so that the model behavior regarding the not-so-well-known physics can be trusted. Data from each well is organized such that it forms one record (one row in a flat file) in the dataset. Each parameter (such as those shown in Fig. 8.61) forms one column in the flat file. The strategy includes dividing the dataset into three separate segments. The first segment known as the "training dataset" will be used to train and build the data-driven predictive model. This segment of the dataset must include all that we need to teach and train the data-driven predictive model. Usually about 70–80 % of the entire dataset (70–80 % of the wells) is dedicated to training the data-driven predictive model.

The other two segments of the dataset that are separated at the beginning of the process are called calibration and validation datasets. The 30–20 % of the wells that are allocated to these two datasets is divided into two equal (and sometimes not equal) segments. These segments of the dataset will not be used to train the predictive model; rather they are used to serve other purposes. The calibration dataset is used to (a) make sure that the data-driven predictive model is not over-trained, and (b) make sure that the best trained predictive model (a model that has learned maximum information content of the dataset) is saved. An over-trained data-driven model actually memorizes the dataset and performs very well in correlating the input parameters to the model output for the training dataset, but has very little technical and scientific value since it is merely a curve fitting exercise (very much like a Decline Curve Analysis) and lacks any kind of verifiable predictive capabilities. The calibration dataset plays the role of a watch dog that oversees the training process and makes sure that the best possible data-driven predictive model is trained.

The third segment of the dataset is the validation (verification) dataset. This segment of the data that is randomly selected and put aside from the beginning of the modeling process plays no role, what-so-ever, in building (training and calibration) of the predictive model. This segment of the dataset is to test the robustness and the validity of the model and see if the data-driven predictive model has learned well-enough to be able to calculate the productivity of shale wells (model output) as a function of all the parameters that were used as the input to the model (Fig. 8.61). The calibration and validation datasets are also known as blind records (blind wells) that validate the goodness and robustness of the predictive model. The model is judged based on its performance on the blind wells not the training dataset.

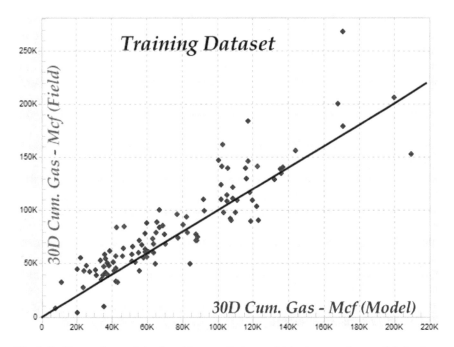

Fig. 8.62 The goodness of the data-driven predictive model is shown on how well it has been trained

In the data-driven predictive model developed for this Marcellus shale asset, 80 % of the wells were used to train the model while the remaining 20 % of the wells were left for calibration and validation purposes. Figure 8.62 shows the results of predictive modeling for the training set and Fig. 8.63 shows the results of the model performance on the blind (calibration and validation) datasets. The training dataset has a $R^2 = 0.76$ and a correlation coefficient of 0.89, while the R^2 and correlation coefficient for calibration dataset are 0.71 and 0.88 and for validation dataset are 0.75 and 0.91, respectively.

These statistics on the model's predictive capabilities are not very impressive, but acceptable for such complex behavior. As we mentioned before, all the data-driven predictive model needs to achieve is a better result than other techniques that have been used to model production from shale. And the results of this model are far better than any other model that can be developed using the traditional methods. It must be noted that the Marcellus shale predictive model that is presented in this chapter is the first data driven models built for shale (several years ago). Since then the author has been involved in building a large number of data driven predictive models for several other shale assets throughout the United Staes (multiple models for Marcellus and Eagle Ford shale, as well as for Utica, Niobrara, and Bakken shale). The Shale Analytics technology for developing data driven predictive model is far more advance today than it was when the model presented here was developed.

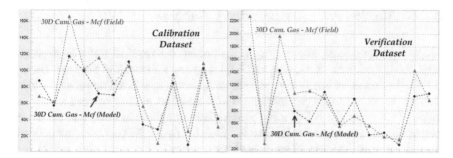

Fig. 8.63 Results of the data-driven predictive model tested against blind wells that were not used to train the model

Once the predictive model is trained, calibrated and validated for accuracy, it can be used to perform analysis. The analyses performed using the predictive model include single and combinatorial sensitivity analyses as well as uncertainty analysis using Monte Carlo simulation technique. Furthermore, type curves can be developed based on the predictive model for any well, groups of wells or the entire field, for any of the combinations of parameters that have been used in the development of the predictive model.

8.7   Sensitivity Analysis

Now that a data-driven predictive model has been developed (trained and calibrated) and validated, we can query the model in order to better understand its characteristics. This can be accomplished through a series of sensitivity analyses. During these analyses, the model behavior is extensively studied in order to understand how the impact of different well, reservoir, and completion-related parameters have been incorporated in the model. Please remember that since no deterministic formulation was used during the construction of the predictive model it is not easy to quickly realize the model behavior under different circumstances. From an engineering point of view, in order to build confidence on the predictive capabilities of a model, we need to understand whether the model makes physical and geological sense.

This goes against the superficial notion that has been circulating around some groups that view data-driven predictive models as black-boxes. Those that make such mistakes usually mistake data-driven predictive models with statistical curve fits that follow no physical and/or geological reasoning. On the contrary, data-driven predictive models that are developed and presented throughout this book are engineering tools (honoring the causality of developed correlations) that follow the principles that make an engineer comfortable with using such models. The next couple of sections in this chapter examine the single and multiple parameters sensitivity of this data-driven predictive model.

8.7.1   Single-Parameter Sensitivity Analysis

Single-Parameter sensitivity analysis is performed on individual wells. Each plot in Figs. 8.64, 8.65, 8.66, 8.67, 8.68 and 8.69 represents one well in the field. During the Single-Parameter sensitivity analysis, parameters are selected one at a time to be studied. While all other parameters are kept constant at their original value, the value of the parameter being analyzed is varied throughout its range and the model output (30 days cum gas production) is calculated (using the predictive model) and plotted for each variation. This procedure is repeated for every well in the asst. Figure 8.64 shows the result of single-parameter sensitivity analysis for porosity. In this figure, the red-cross on each plot identifies the actual production value of the well and the corresponding porosity value used for that particular well. In Fig. 8.64, (similar to the other figures that show the single-parameter sensitivity analysis) three wells are shown for each parameter.

It has to be noted that since all the parameters (other than the one being analyzed) is kept constant at their original value, these analyses (on an individual well basis) may look a bit strange from time to time. This is the result of an intuitively linear analysis (single-parameter sensitivity analysis) performed on a highly non-linear system.

In this figure (Fig. 8.64), the *y*-axis in all three plots, is the model output (30 days cum gas production) while the *x*-axis is the porosity. In these plots, we can see the changes in 30 days cum gas production as a function of porosity in the Marcellus shale. In all three examples shown in this figure, 30 days cum gas production increases as the porosity increases. It should also be mentioned here that in the original dataset reservoir characteristics (including porosity) were made available for Lower Marcellus and Upper Marcellus separately. These values were combined in order to arrive at the reservoir characteristics (including porosity) used in as Marcellus shale, in these analyses.

Figure 8.65 shows single-parameter sensitivity analysis for four other Marcellus shale reservoir characteristics in the dataset, namely, Gross Thickness, Net to Gross

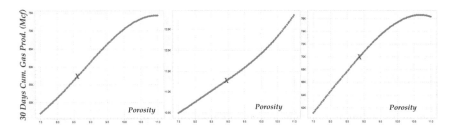

Fig. 8.64 Single-parameter sensitivity analysis (reservoir characteristics)—porosity—examples shown for three wells in the asset

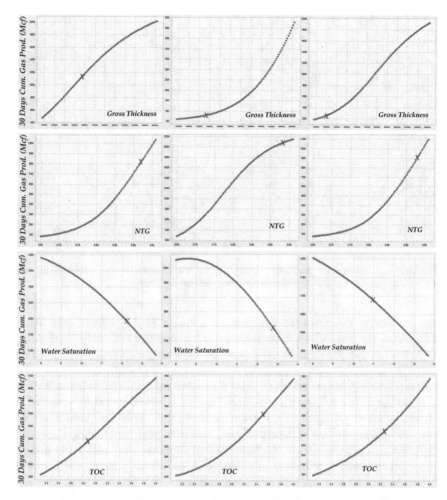

Fig. 8.65 Single-parameter sensitivity analysis (reservoir characteristics)—gross thickness, NTG, Sw, and TOC—examples shown for three wells in the asset

ratio, Water Saturation, and Total Organic Content (TOC). Examples of three wells are shown for each parameter and in each of the plots a red "X" indicates the actual value of the parameter for each well and the corresponding 30 days cum gas production. The trends that are shown in these examples follow the expected trends. The top row of the plots shows that as the Gross Thickness of the Marcellus shale increases so does the 30 days cum gas production. The second row of the plots shows that as the Net to Gross ratio of the Marcellus shale increases so does the 30 days cum gas production. The third row of the plots shows that the 30 days cum

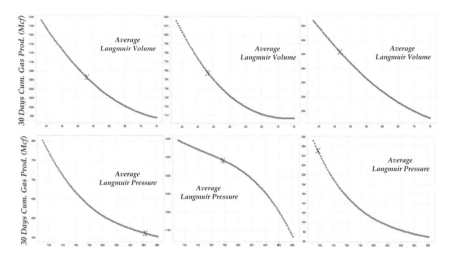

Fig. 8.66 Single-parameter sensitivity analysis (reservoir characteristics)—Langmuir pressure and volume constants—examples shown for three wells in the asset

gas production decreases as function of increase in Marcellus shale Water Saturation (which translates in an increase in hydrocarbon saturating), and finally the bottom row of the plots shows that as the Total Organic Content (TOC) of the Marcellus shale increases so does the 30 days cum gas production.

It is interesting to note that in all these trends that are shown here almost 90 % of them do not follow a linear change. Some sort of nonlinearity can be observed in the overwhelming majority of the trends of all the wells and almost all the parameters being studied in these analyses. This is an important finding since most engineers and geoscientists in the industry that somehow deal with shale instinctively believe such nonlinearity in behavior exists but have never had a tool or technique to shed light on such behavior, until now.

Figure 8.66 shows single-parameter sensitivity analysis performed for the final two reservoir characteristics parameters in the dataset, namely, Average Langmuir Pressure Constant and Volume Constant in Marcellus shale. The trend of 30 days cum gas production as a function of changes in these parameters as shown in this figure and as expected in both cases decrease in 30 days cum gas production is observed, although nonlinearly that is different from well to well, as the Average Langmuir Pressure and Volume Constants decreases in magnitude.

Figure 8.67 shows single-parameter sensitivity analysis for the parameters that have been categorized as completion in this study. These parameters are Perforated and Stimulated Lateral Length, Shot Density, and the number of Clusters per Stage. As expected, increasing Perforated and Stimulated Lateral Length causes an increase in 30 days cum gas production, though in a nonlinear fashion. The

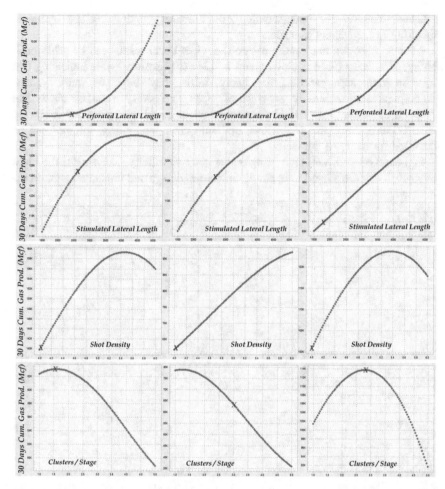

Fig. 8.67 Single-parameter sensitivity analysis (well completion)—perforated and stimulated lateral length, shot density, and number of Clusters per Stage—examples shown for three wells in the asset

Figure shows that better 30 days cum gas production can be achieved with higher Shot Density while smaller numbers of Clusters per Stage seem to be preferable, while some wells show that an optimum number of Clusters per Stage can be identified.

The common theme among all these parameters is the nonlinear nature of the plots in these two figures. These plots relate the 30 days cum gas production with all the hydraulic fracturing design and implementation parameters in the Marcellus shale, on a well by well basis. This nonlinear behavior underlines the complex nature of these relationships and demonstrates the inadequacy of using simple statistical analyses for understanding and ultimately optimizing the hydraulic fracturing in the Marcellus shale.

Average injection rates and pressure show that better results can be achieved if the injection takes place at slower rates and lower pressures, while indicating that injection rate and pressure for many wells may have optimum values. Single-parameter sensitivity analysis of clean and slurry volumes show that although in general lower fluid volumes seem to be advantageous for achieving better 30 days cum gas productions, optimum clean, and slurry volumes can be identified for many of the wells.

Figures 8.68 and 8.69 show single-parameter sensitivity analysis for the eight parameters that are related to hydraulic fracturing design and implementation in the

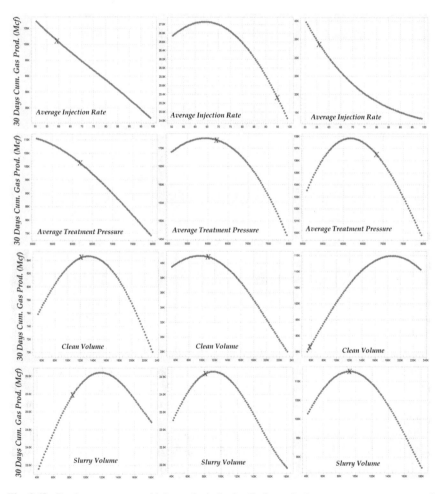

Fig. 8.68 Single-parameter sensitivity analysis (hydraulic fracturing)—average treatment rate and pressure, clean and slurry volumes—examples shown for three wells in the asset

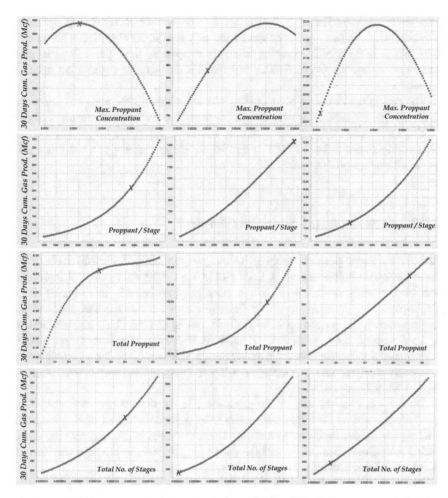

Fig. 8.69 Single-parameter sensitivity analysis (hydraulic fracturing)—maximum proppant concentration, proppant per stage, total proppant, and total number of stages—examples shown for three wells in the asset

Marcellus shale. Figure 8.68 shows the analyses for the Average Treatment Rate and Pressure as well as amount of Clean and Slurry Volumes, while Fig. 8.68 shows the analyses for the Maximum Proppant Concentration, Proppant per Stage, Total Proppant, and Total Number of Stages.

In the top row of the Fig. 8.68, single-parameter sensitivity analysis is shown for Maximum Proppant Concentration. These analyses show that an optimum Maximum Proppant Concentration can be identified for a large number of wells.

Maximum Proppant Concentration refers to a design parameter during hydraulic fracturing that the frac engineers should use as a goal or an objective. The process starts at the low concentration that is slowly ramped up throughout the operation and therefore is not a constant value that remains the same throughout the fracking operation.

The two middle rows of plots in Fig. 8.68 show that higher proppant injections (both in a per stage basis as well as total for the entire well) usually result in higher production, acknowledging the nonlinear nature of this relationship. The bottom plots in this figure points to the fact that higher number of stages usually result in wells with better production.

8.7.2 Combinatorial Sensitivity Analysis

Just like the single-parameter sensitivity analysis, combinatorial sensitivity analysis is performed on individual wells. Each plot in Figs. 8.70, 8.71 and 8.72 represents one of the wells in the field. When sensitivity analysis is investigated for a single-parameter, it results in a two-dimensional plot where the sensitivity of the target parameter is gauged against the model output (30 days cum gas production) as seen in Figs. 8.64, 8.65, 8.66, 8.67, 8.68 and 8.69. If sensitivity analysis is investigated for two parameters, simultaneously, the result is called a combinatorial sensitivity analysis and is displayed using a three-dimensional plot as seen in Figs. 8.70, 8.71 and 8.72.

During this process, two of the parameters in the dataset (for each individual well) are selected as the target parameters to be studied. While all other parameters are kept constant at their corresponding values, the value of the target parameters is varied throughout their ranges and the model output (30 days cum gas production) is calculated and plotted for each variation.

Figure 8.70 shows the results of combinatorial sensitivity analysis performed on six wells. In this figure, examples of combinatorial sensitivity analyses that have been performed on reservoir characteristics are displayed. The parameters that have been analyzed are Porosity and Gross Thickness (top two plots), Water Saturation and NTG (middle two plots), and TOC and Gross Thickness (bottom two plots). In each of the plots, the presented surface shows the changes in 30 Days Cumulative Gas Production (the colors simply represent different 30 Days Cumulative Gas Production values). As a function of the two parameters that are being analyzed. The "White Cross" on each of the plots represents the actual value of the 30 Days Cumulative Gas Production that was achieved in the field and the corresponding reservoir characteristics for each well.

Figure 8.71 shows the results of combinatorial sensitivity analysis performed on six wells. In this figure, examples of combinatorial sensitivity analyses that have

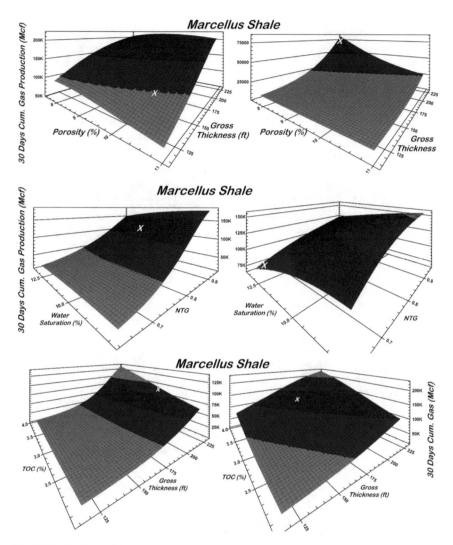

Fig. 8.70 Combinatorial sensitivity analysis (reservoir characteristics)—porosity and gross thickness—water saturation and NTG—TOC and gross thickness

been performed on completion parameters are displayed. The parameters that have been analyzed are Clusters per Stage and Perforated Lateral Length (top two plots), Shot Density and Stimulated Lateral Length (middle two plots), and Shot Density and Clusters per Stage (bottom two plots).

Figure 8.72 shows the results of combinatorial sensitivity analysis performed on six wells. In this figure, examples of combinatorial sensitivity analyses that have

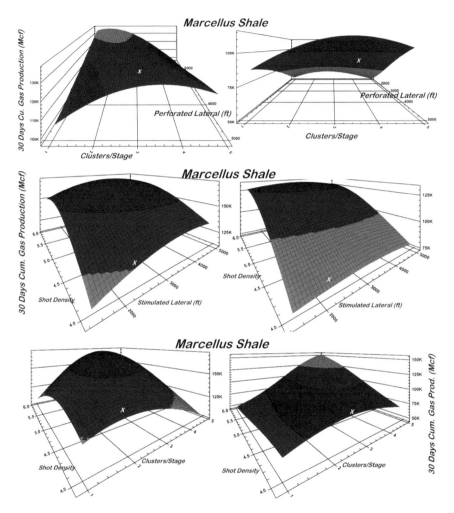

Fig. 8.71 Combinatorial sensitivity analysis (completion)—clusters/stage and perforated lateral length—shot density and stimulated lateral length—shot density and clusters/stage

been performed on hydraulic fracturing parameters are displayed. The parameters that have been analyzed are Proppant per Stage and Treatment Pressure (top two plots), Slurry Volume and Total Proppant (middle two plots), and Maximum Proppant Concentration and Total Number of Stages (bottom two plots).

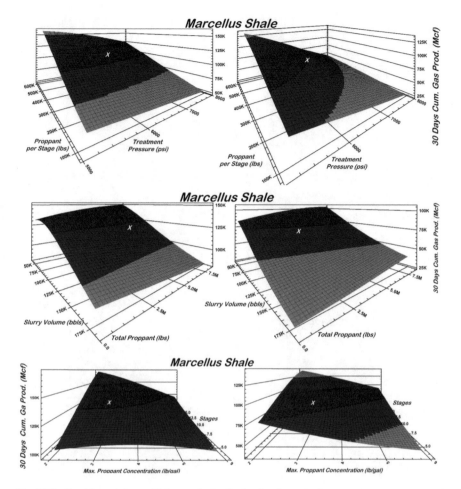

Fig. 8.72 Combinatorial sensitivity analysis (hydraulic fracturing)—proppant per stage and treatment pressure—slurry volume and total proppant—maximum proppant concentration and total number of stages

8.8   Generating Type Curves

Upon successful development of the predictive model, type curves can be generated to assist operators during the decision making process on where to place the next well (or which planned wells should get priority for drilling) and how to complete and stimulate it. Type curves can be generated for individual wells, for groups of wells (corresponding to a specific area in the asset) and for the entire field. In type curves, the y-axis is the model output (in this case, 30 day cum gas production). The x-axis should be selected from one of the input parameters and the generated curves represent a third parameter.

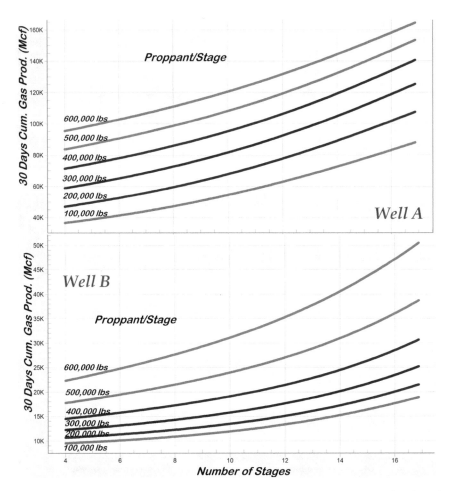

Fig. 8.73 Type curves for two wells showing changes in 30 days Cum. gas as a function of number of stages and different proppant/stage

Figure 8.73 shows two sets of 30 days cum gas production type curves for Proppant per Stage (curves) as a function of number of Stages (x-axis) for two locations (identified by two wells) in the dataset. Similar type curves can be developed for all the wells in the field (Figs. 8.74, 8.75, 8.76, 8.77, 8.78, 8.79 and 8.80). Since these type curves have been generated for individual wells, they can be interpreted as type curves that represent well behavior in specific locations of the field. To generate type curves for other locations in the field (loactions that are targeted for new wells), once can simply modify the reservoir characteristics to represent the new location and re-generate the type curves.

One important issue that needs to be emphasized here is the well-behaved nature of the curves that are displayed in Figs. 8.73, 8.74, 8.75, 8.76, 8.77, 8.78, 8.79 and 8.80. Type curves are well-known in our industry. Almost all the type curves that

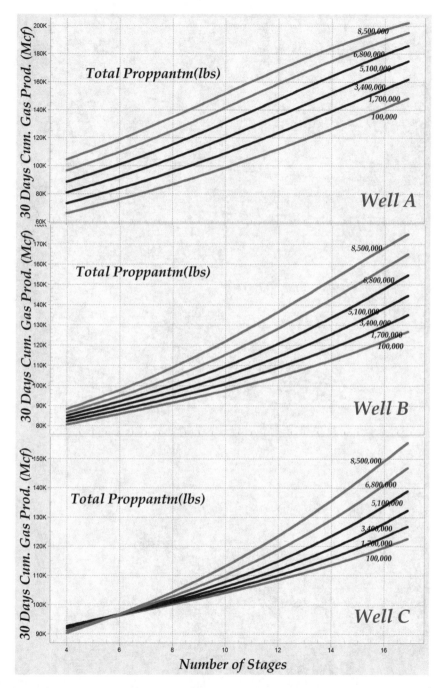

Fig. 8.74 Type curves for three wells showing changes in 30 Days Cum. Gas as a function of number of stages and different total proppant

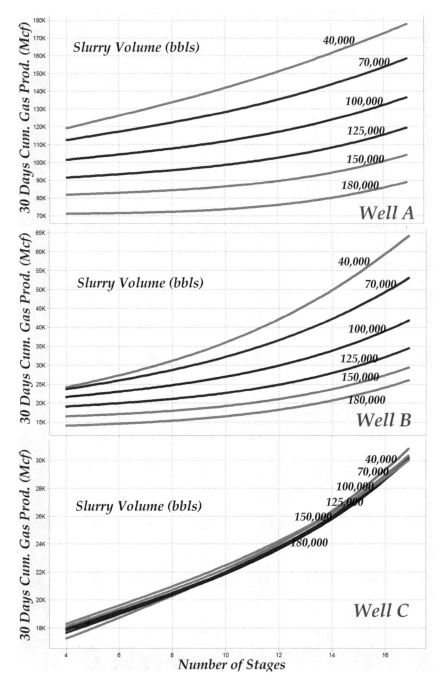

Fig. 8.75 Type curves for three wells showing changes in 30 days Cum. gas as a function of number of stages and different slurry volumes

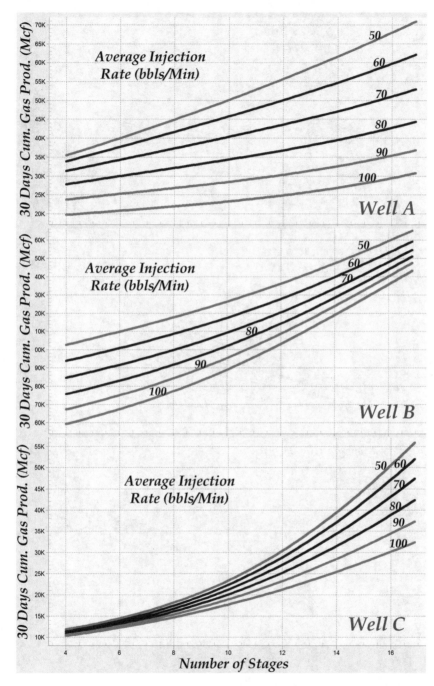

Fig. 8.76 Type curves for three wells showing changes in 30 days Cum. gas as a function of number of stages and different average injection rates

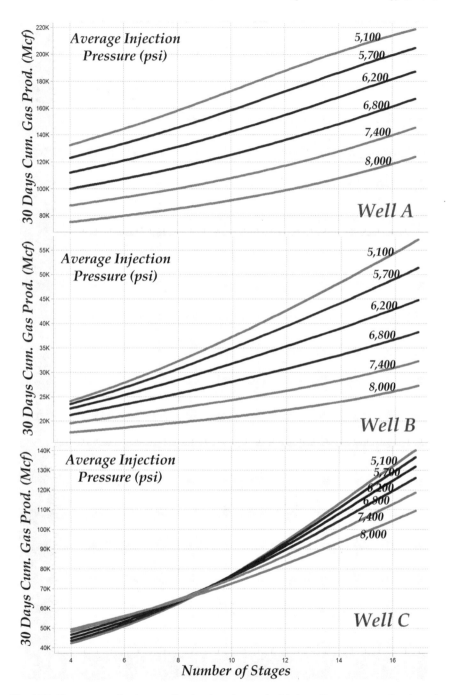

Fig. 8.77 Type curves for three wells showing changes in 30 days Cum. gas as a function of number of stages and different average injection pressure

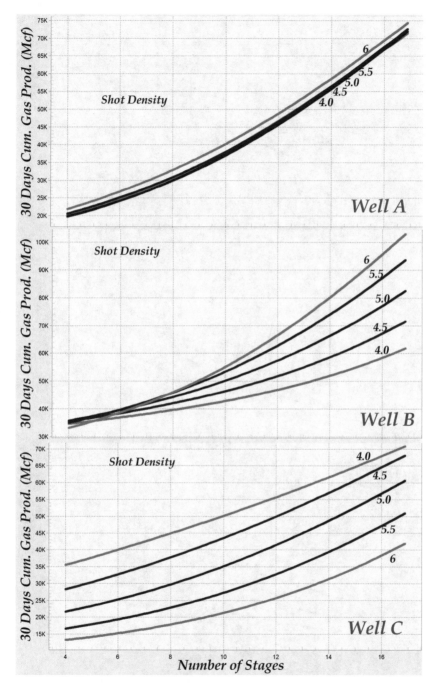

Fig. 8.78 Type curves for three wells showing changes in 30 days Cum. gas as a function of number of stages and different shot density

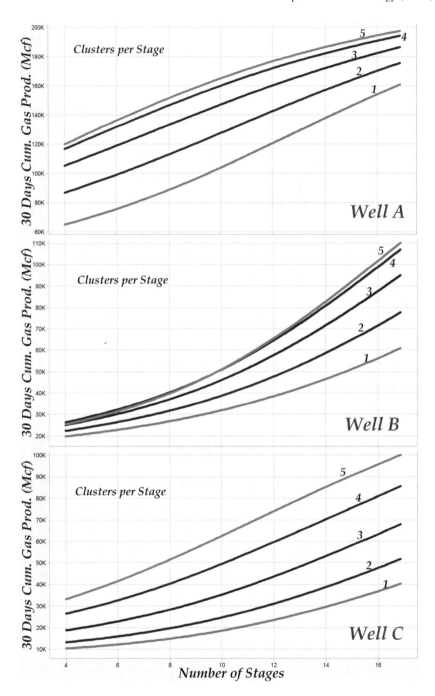

Fig. 8.79 Type curves for three wells showing changes in 30 days Cum. gas as a function of number of stages and different cluster/stage

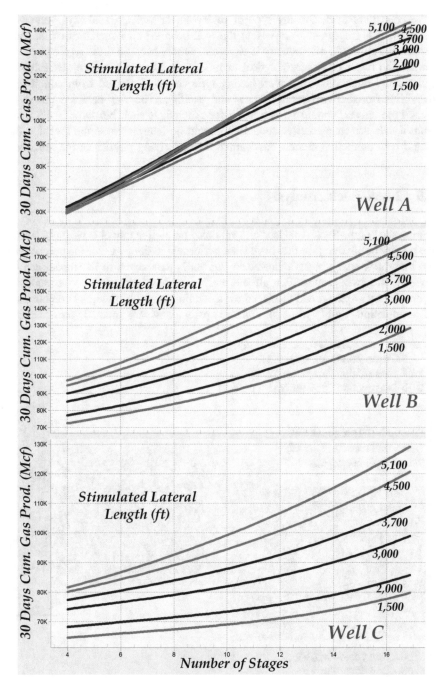

Fig. 8.80 Type curves for three wells showing changes in 30 days Cum. gas as a function of number of stages and different stimulated lateral length

have been published, for any type of reservoirs or wells, are always very well behaved. It is important to note that the well-behaved characteristics of all published type curves in our industry is expected, simply because these type curves are always the result of either analytical or numerical solution of some deterministic (governing) equations. However, having type curves generated from completely data-driven predictive models is new and exciting. The well-behaved nature of these type curves, should not be taken for granted. If such behavior is observed from a data-driven predictive models it should be interpreted as the fact that the data-driven predictive models has learned the underlying physics of the process.

8.9   Look-Back Analysis

During the "Look-Back Analysis" the validated predictive model is used to identify the quality of the frac jobs that have been performed in the past. For the purposes of the look-back analysis, as indicated in Fig. 8.81, parameters in the model can be divided into two groups. For all the wells that have already been completed, hydraulically fractured and produced for a certain amount of time, the combination of completion and hydraulic fracture parameters are named the "Design

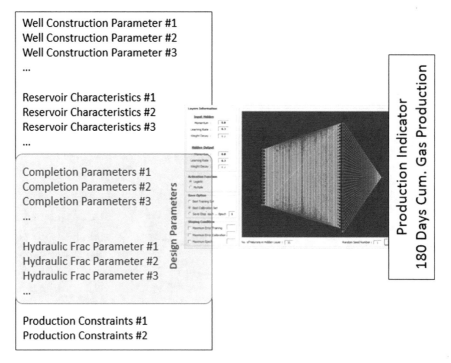

Fig. 8.81 Parameters involved in development of the data-driven predictive model. Design (completion and hydraulic frac) parameters are identified in the figure

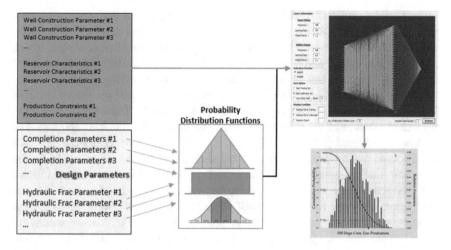

Fig. 8.82 Schematic diagram demonstrating the look-back analysis

Parameters." When looking back at the well's performance after it was drilled, a combination of completion and hydraulic fracture parameters as well as production constraints are the parameters that controlled the production (degree of productivity of the well).

The questions that is going to be addressed during this analysis are:

1. "What was the production potential of this well, given the location (reservoir characteristics) and the well construction?"
2. "How' much of the potential was realized with the specific completion and hydraulically fracturing practices performed on this well?"
3. "What percent of the wells in this asset are producing as expected, better than expected, or worse than expected?"

In order to find reliable answers to the above questions, the following algorithm is devised. This algorithm uses Monte Carlo Simulation technique in order to accomplish its task. The flow chart described in Fig. 8.82 is performed for each well in the asset, individually. During this process, the input parameters into the predictive model are divided into two groups. Group one include well construction parameters, reservoir characteristics, and operational constraints and group two are design parameters including completion and frac parameters. The objective is to find out the production potential of each well based on its well construction, reservoir characteristics and operational constraints. For each individual well, this is accomplished by keeping group one parameters constant, while modifying group two (design) parameters.

During the Monte Carlo Simulation, each time the predictive model is executed, group one inputs are kept constant while group two parameters (Design - completion or hydraulic fracture parameters) are modified to a new combination. Each time the combination of the group two parameters are selected randomly based on the assigned a probability distribution function as shown in Fig. 8.82 and a new production indicator (for example, 180 days cumulative gas production) is calculated. This way, we learn what would this well be capable of producing if it was completed and stimulated with a new set of design parameters. The probability distribution functions can be assigned based on the target of the analysis.

During this Monte Carlo Simulation analysis, the data-driven predictive model as the objective function of the process, is executed thousands of times for each well to produce the output of the model after each execution. This results in thousands of production indicator (for each well), each time with the same set of reservoir, well construction, and operational characteristics but different completion and hydraulic fracturing (design) parameters. The thousands of values of the production indicator that has been generated in this fashion represent thousands of completion and stimulation practices for a given well. These collection of results represent the complete range of possible production values from the given well. These results are plotted in the form of a probability distribution function from which P10, P50, and P90 are calculated (please see examples presented in Fig. 8.83).

The collection of the model outputs represents the potential productivity of this well. The minimum, the maximum as well as the average production that could have been expected from a given well that is constructed in a certain way and is producing from a shale formation with certain characteristics and have been subjected to certain operational constraints. Given the probability distribution that is generated for each well as the outcome of the analysis, we can generate a series of judgments regarding the quality of the frac job as a function of well productivity. We recommend the following rules for the analyses regarding the quality of the frac job:

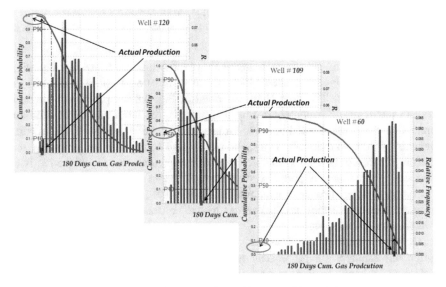

Fig. 8.83 Look-back analyses performed on three different well in a shale asset

1. If the well produces at production values that is *above P20* (of its potential), the quality of the frac job is identified as "***Excellent***,"
2. If the well produces at production values that is *between P40 to P20* (of its potential), the quality of the frac job is identified as "***Better than Expected***,"
3. If the well produces at production values that is *between P60 to P40* (of its potential), the quality of the frac job is identified as "***As Expected***,"
4. If the well produces at production values that is *between P80 to P60* (of its potential), the quality of the frac job is identified as "***Worse than Expected***,"
5. If the well produces at production values that is *below P80* (of its potential), the quality of the frac job is identified as "***Poor***."

Once the above rules are in place, then the actual value of the Production Indicator of the well is projected on to the probability distribution function and the quality of the completion is judged based on the range within which the actual production is fallen. Figure 8.83 shows the results of "Look-Back" analysis performed on three shale wells completed in Marcellus Shale in Pennsylvania. In this figure the completion of Well #120 (top-left) is classified as "Poor" since the actual (historical) productivity (180 days of cumulative production) of this well falls on P-97. The completion Well #109 (middle) is classified as "As Expected" since the actual (historical) productivity (180 days of cumulative production) of this well falls on P-52, and finally the completion of Well #60 (bottom-right) is classified as "Excellent" since the actual (historical) productivity (180 days of cumulative production) of this well fall on P-5.

Once this analysis is performed on all the wells in an asset, the results for all the wells in the asset can be compiled in order to identify the number (and percentages) of the wells that fall into each of the five categories of completion quality. Figure 8.84 shows the final results of the "Look-Back" analysis for a Utica shale asset in Ohio. In this asset, based on the Look-Back analysis, 43 % of the wells

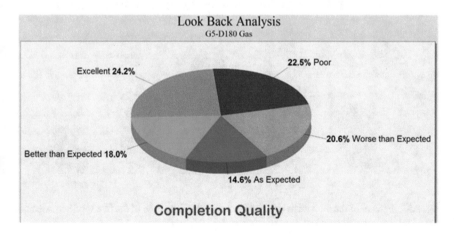

Fig. 8.84 Final result of a complete "look-back" analysis performed on a utical shale asset

were completed and fracked poorly and have resulted in worse than expected (meaning that more hydrocarbon could have and should have been produce from these wells) productivity, while 43 % of the wells were completed and fracked such that better than average production has been recovered. About 15 % of the wells were fracked as expected and hydrocarbon production from these wells was in the range of expected values. Of course, operators may state that their overall expectations in all their operations are to produce above average from every well in their asset. Once Shale Analytics is used in a company, the primary goal can be to improve the overall expectations from the completion.

8.10   Evaluating Service Companies' Performance

Operators usually use multiple service companies for completion. It would be of interest to have a yardstick against which the operators can measure the quality of the completion and frac jobs performed by each service company. By keeping the reservoir characteristics constant for each well during the analysis that was presented in Sect. 8.9 the impact of completion and stimulation parameters on the production was isolated. Therefore, this would be an appropriate way to compare the quality of completion and stimulation practices of different service companies.

In a recent project performed on a Marcellus shale asset that included more than 230 wells, this technology was used to compare the completion and hydraulic fracturing practices of five service companies that had been used in this asset. Figure 8.85 shows the percent of the wells in the asset that has been completed and hydraulically fractured by each of the five service companies. Figure 8.86 demonstrate the average productivity (180 days cumulative gas production) of

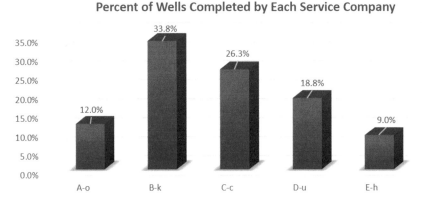

Fig. 8.85 Percent of the total wells that had been completed by each of the five service companies

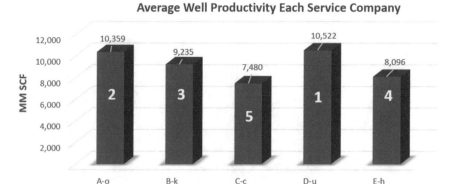

Fig. 8.86 Average 180 days of cumulative gas production for each well that the service company has completed

wells that has been completed by each of the service companies. Name of companies are identifies by alphabetical letters such as "A-o," "B-k," "C-c," "D-u," and "E-h."

As it can be seen from Figs. 8.85 and 8.86, Service Company "E-h" has completed the least number of wells (9 % of all the wells in the asset) with an average 180 days cumulative gas production of about 8 million SCF per well while Service Company "B-k" has completed the most number of well (34 % of all the wells in the asset) with an average 180 days cumulative gas production of about 9.2 million SCF per well. Service Company "D-u" has completed the 20 % of the most productive wells (on average) in this asset with an average 180 days cumulative gas production of about 10.5 million SCF per well and Service Company "C-c" has completed the 26 % of the least productive wells (on average) in this asset with an average 180 days cumulative gas production of about 7.5 million SCF per well.

"Px" denotes the location of the production indicator value on the probability distribution function generated by the Monte Carlo Simulation for each well. For example, at x = 50, the Px = P50 denotes the average value of the production indicator on the probability distribution function generated by a Monte Carlo Simulation for a given well (when the value of "x" in Px, is small, it indicates better productivity). The "Px" in Fig. 8.87 is the average for all the wells completed by each of the service companies. This figure indicates that Service Company "A-o" (with Px = P24) has had the best performance followed by Service Company "E-h" (with Px = P35), while Service Company "D-u" has performed the poorest of all the five service companies (with Px = P57) in this Marcellus Shale asset.

Figures 8.88 and 8.89 show actual production numbers. In Fig. 8.88 total production for each of the service companies below P10 is shown. Let us explain how the numbers shown in this figure have been calculated. For each well, the actual production (180 days cumulative gas production) is subtracted from the P-10

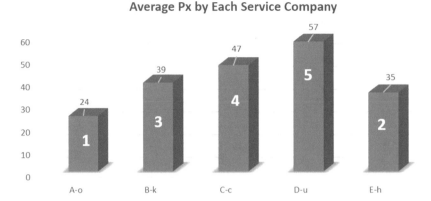

Fig. 8.87 Average Px of all the wells that have been completed and stimulated by each of the service companies

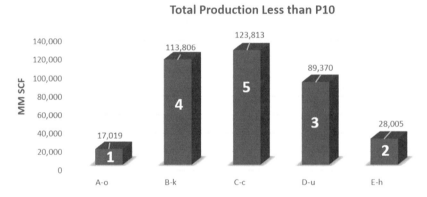

Fig. 8.88 Total production less than P10 values for the wells completed by each service company

production that had been calculated for the well using the Monte Carlo Simulation procedure explained in Sect. 8.9. Then this difference was summed over all the wells that had been completed by each service company.

The numbers displayed in Fig. 8.89 is calculated by dividing the total production values (shown in Fig. 8.88) by the number of wells completed by each service company to arrive at an average value. Comparing the numbers shown in Figs. 8.85, 8.86, 8.87, 8.88 and 8.89, one can conclude that completions and stimulations performed by the Service Company "A-o" is of the highest quality amongst these five service companies, followed by the Service Company "E-h." Furthermore, these figures show that Service Company "D-u" has demonstrated to

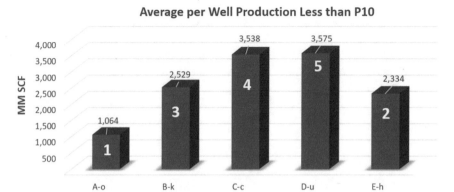

Fig. 8.89 Average per well production less than P10 values for the wells completed by each service company

have the poorest quality completion and stimulation practices in this group of service companies.

Chapter 9
Shale Numerical Simulation and Smart Proxy

The authors' believe in the lack of effectiveness of numerical simulation in modeling production from shale does not stem from a purely theoretical point of view. We have been intimately involved in such activities for years and have learned that when it comes to using numerical reservoir simulation to model production from shale wells, regardless of how detail and meticulous the approach, the results leave much to be desired and have little to do with the realities on the ground. We have concluded that given the lack of reality-based constraints on the model, one can get whatever results she/he wishes from the numerical simulation models. Many experienced reservoir modelers in some of the top international companies have come to the same conclusions and some have even publicly expressed such opinions.

In this chapter we have decided to present one of the many studies that have been performed using numerical reservoir simulation that has benefited from a new data-driven analytics technology called "Smart Proxy".

9.1 Numerical Simulation of Production from Shale Wells

Numerical simulation of production from shale has been put together by combining several techniques that had been developed in the past several decades to model natural fractured reservoirs, and to model storage and transport phenomena driven by adsorption and concentration gradient above and beyond conventional mechanisms.

This Chapter is Co-Authored by: Dr. Amir Masoud Kalantari, University of Kansas.

© Springer International Publishing AG 2017
S.D. Mohaghegh, *Shale Analytics*, DOI 10.1007/978-3-319-48753-3_9

9.1.1   Discrete Natural Fracture Modeling

Modeling gas-matrix-fracture phenomena and developing reliable and fast reservoir assessment techniques play very important role to fulfill the fast development plan for shale assets. Numerical reservoir simulation have been used extensively by many operators and service companies during the past several years to perform production optimization and reserve estimation in shale formation.

Dual continuum models are the conventional method for simulating fractured systems and are widely used in the industry (Fig. 9.1). Barenblatt et al. [68] and Warren and Root [69] first developed mathematical models describing the fluid flow in dual-porosity media. These models assume single-phase fluid in pseudo-steady state, transfer from matrix to fracture. The Single-phase, Warren and Root approach was later extended to the multiphase flow [70–73] and the dual-porosity simulators were developed. These new simulators were capable of modeling unsteady state or transient fluid transfer between matrix and fracture. In this model, naturally fractured media was represented as a set of uniform matrix blocks and fractures as demonstrated in Fig. 9.1.

Later models were developed to accommodate dual permeability models, which allow for matrix to matrix and fracture-to-fracture flow and can be used to model transient gas production from hydraulic fractures in shale gas reservoirs [74, 75]. Dual-Mechanism approach (Darcy flow and Fickian diffusion occur parallel in matrix) was introduced to characterize the gas flow in coal or shale formations through the dynamic gas slippage factor [76].

The most common technique for modeling Discrete Natural Fracture (DNF) network is to generate them stochastically. Using Borehole Image Logs (e.g. FMI), some of the initial characteristics of the DNF can be estimated and used for their stochastic generations. Parameters such as fracture point's dip angle, dip azimuth, averages for fracture length, aperture, density of center points are among

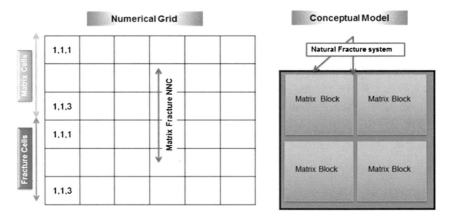

Fig. 9.1 Dual-porosity numerical and conceptual model

those that need to be provided (guessed or estimated) so that the stochastic algorithms can generate a Discrete Natural Fracture (DFN) network.

FMI-interpreted fractures are subject to easier opening by hydraulic fracturing than virgin shale rock. From the practical point of view, all types of interpreted fractures can be considered as constituting a "natural" fracture network that partially dominates hydraulic fracture network intensity and distribution. Additional important factors, such as in situ stress field and geomechanical properties, have great impact on generation of complex fracture systems.

A generated DFN model is not useful until upscaled into the required grid properties that can be used for performing dual continuum (dual-porosity or dual-porosity/dual permeability) flow simulation. The new grid properties are fracture permeability (either diagonal or full tensor), fracture porosity and shape factor (matrix-fracture transfer function). The final properties for each grid will be obtain during the history matching process. Fracture porosity is calculated using the following equation:

Formulation to calculate fracture porosity.

$$\emptyset_f = \text{Total fracture area} * \text{Aperture}/\text{Volume of cell} \qquad (9.1)$$

The numerically derived expression for sigma (shape) factor in terms of fracture spacing (matrix block size) in $i, j,$ and k grid coordinate directions of given cell is as follows:

Formulation for calculation of shape factor.

$$\sigma = 4 * \left(\frac{1}{L_i^2} + \frac{1}{L_j^2} + \frac{1}{L_k^2} \right) \qquad (9.2)$$

9.1.2 Modeling the Induced Fractures

Apart from understanding the complex physics in shale reservoirs and trying to model it, the economic production of these (mostly) completion–driven reservoirs highly depend on the quality of the created massive multiple cluster hydraulic fractures and their interaction with the rock fabric. However, the modeling of shale reservoir becomes even more complicated with the existence of such complex fracture network.

The hydraulic fracturing field data (e.g. Proppant amount/size, slurry volume, injection pressure and rate, etc.) cannot be incorporate into the numerical models, directly. For this reason, analytical and semi-analytical techniques can be used to model hydraulic fractures initiation and propagation and calculate (estimate) the corresponding properties by taking into consideration the state of stress and geomechanics that can be used in a form of grid property in the reservoir simulation.

Explicit Hydraulic Fracture (EHF) modeling method, and the Stimulated Reservoir Volume (SRV) are two main approaches to incorporate the calculated hydraulic fracture characteristics into the numerical simulation, and each requires a proper gridding technique. Between these different approaches, EHF is the most detail and complex way to model "the effect" of hydraulic fracturing on shale production in numerical simulation. Despite complexity, this approach attempts to implicitly honor the hydraulic fracturing data at the cluster/stage level and incorporate such information in the simulation model.

Model setup for the EHF technique is long and laborious and its implementation is computationally expensive, such that it becomes difficult or in some cases impractical to model beyond a single pad. By reviewing the numerical reservoir simulation modeling efforts in the literature with focus on shale reservoirs, it can be noticed that almost all published studies are concentrated on the simulation of production from single shale well [29, 77–79]. This confirms the complexity and computationally prohibitive use of this technique, especially in the case of performing simulation at multiple pad or full-field level.

Stimulated Reservoir Volume (SRV) is another/alternative technique for modeling the impact of created and propagated hydraulic fractures in shale formation. This approach was initially proposed when the extension and orientation of created hydraulic fractures were calculated (estimated) based on microcosmic events. This technique, makes the simulation and modeling process simpler and much faster, compared to the EHF method, by modeling the hydraulic fracture volumetrically with enhanced permeability region around the wellbore. Numerous published papers addressed the geometry and shape of the SRV created around horizontal wellbores in shale reservoirs using microseismic monitoring techniques with simple diagnostic plots [80–82]. Most of these efforts are not calibrated quantitatively by mathematical correlations to production data.

On the other hand, in most of the shale simulation studies, microseismic data is not available; therefore this approach can and has been missused to simplify the history matching process by only changing the permeability around the wellbore and without taking into account the hydraulic fracturing field data, which ultimately results in inaccurate forecast, even with achieving good history matching result. Therefore, in order to use the numerical simulation as a tool for shale reservoir management, and necessity of properly incorporating hydraulic fractures by honoring the hydraulic fracturing hard data (measured at the field) in the model while having limited access to microseismic data, the EHF technique is a physically more accurate approach. The only problem is to find a solution to improve the model development and simulation run time.

9.2 Case Study: Marcellus Shale

To numerically model the production from a pad that includes multiple horizontal wells completed in the Marcellus shale in the Southwestern Pennsylvania, an integrated workflow was developed. This workflow, which demonstrates a quantitative platform to model shale gas production through capturing the essential characteristics of shale gas reservoirs is shown in Fig. 9.2.

9.2.1 Geological (Static) Model

Developing a detailed geological model in the vicinity of the well, requires the integration of wellbore measurements (e.g. core data, well logs), which provide important details of the lateral heterogeneity and variation of reservoir and rock properties. Information from 77 Marcellus shale gas wells in the Southwestern Pennsylvania is used to develop the geological model for this study.

Property modeling is the process of filling cells with discrete or continuous properties, by using all available geological information. Sequential Gaussian Simulation (SGS) is used to generate the geostatistical distribution of all the properties including matrix porosity, matrix permeability, net to gross ratio, thickness, TOC and also geomechanical properties such as Bulk Modulus, Shear

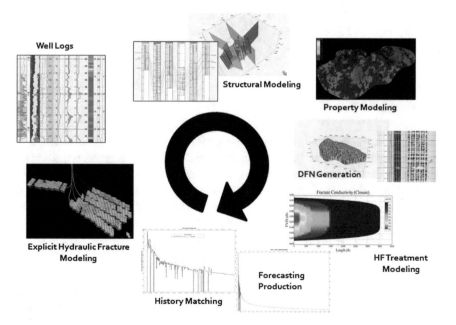

Fig. 9.2 Marcellus shale geomodeling to simulation workflow

Modulus and Young's Modulus, Poisson's ratio and Minimum horizontal stress for all rock types.

Marcellus shale is extremely variable in thickness, ranging from a few feet to more than 250 ft in thickness, and generally becomes thicker to the east. Due to different well trajectory (deviation type), landing targets and completed stages, the pay zone is fully or partially accessible. The Net to Gross ratio ranges from 0.74 to 0.98. Matrix porosity and matrix permeability are changing from 5 to 12.5 % and 0.00018 to 0.0009 md, respectively. In general, Lower Marcellus has higher quality than Upper Marcellus in terms of matrix porosity, matrix permeability, NTG and TOC.

To continue with the workflow, 21 FMI logs are used to model natural fracture distribution in the Marcellus shale. FMI-interpreted fractures are classified, analyzed and two important properties of dip angle and dip azimuth are extracted in several depths for each well. The fracture data points are then used to generate Intensity log, which is used as fracture density volume for the generation of DFN. With proper property drivers, fracture intensity 3D distributions can be achieved, mostly with stochastic simulations. With given well control fracture dips and azimuths input as constants, 2D or 3D properties, and specific fracture geometry specifications, 3D discrete fracture network (DFN) is generated and then upscaled using ODA method.[1]

Calculating hydraulic fracture properties and incorporating in the simulation model is the final step before making a dynamic simulation case. The hydraulic fracture properties (e.g. fracture length, height, porosity and fracture conductivity) for 652 stages and 1893 clusters are calculated explicitly using actual hydraulic fracturing data (hard data) such as proppant amount/size, clean volume, slurry volume, injection pressure/rate, fluid loss, well trajectory, perforation, geomechanical properties and stress.

9.2.2 Dynamic Model

Due to the computational limitations for performing numerical simulation for an entire reservoir, one of the pads (called WVU pad) with six horizontal laterals in the study area is selected to perform the history matching process. A dual-porosity dynamic model, with discretized matrix blocks and with the grid refinement to explicitly represent the hydraulic fracture, consisting of 200,000 grid blocks with three simulation layers in non-refined regions and nine simulation layers in the refined regions is developed.

---

[1]This method primarily relies on the geometry and distribution of fractures in each cell to build permeability tensor. It uses a statistical method based on the number and sizes of the fractures in each cell. It is fast but does not take into account the connectivity of fractures and can therefore underestimate fracture permeability when the intensity is low.

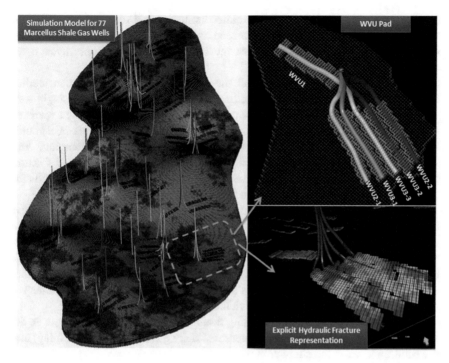

Fig. 9.3 A 3-D view of the simulation model and configuration of the target WVU pad

Logarithmic local grid refinement is performed and each host simulation cell is divided into seven grid blocks laterally and three vertically. The finest grid has 1 ft width and represents the hydraulic fracture in the reservoir simulation model, which possesses the calculated hydraulic fracture characteristics. History matching process is completed for all the WVU pad's laterals based on the 3 years of daily gas production. Figure 9.3 shows the entire study area with 77 horizontal wells and the location of the target WVU pad for history matching.

9.2.3   History Matching

The last step in the development of shale gas simulation model is to adjust the geological model and its parameters such that the simulation model is able to reproduce the gas rate and bottom-hole pressure histories with reasonable accuracy. This history matching is the same as solving an inverse problem, meaning that there is no unique solution and new simulation models can be different from the base geological model.

Generally, in modeling and simulation, the two time-consuming tasks are data gathering and history matching. Shortcuts in gathering data often increases the time needed for history matching because either limited or bad data require additional trial-and-error iterations. Moreover, in case of EHF modeling long simulation run time is a serious issue, which makes the history matching process even more complicated.

It should be noted that, almost all the existing numerical simulation studies conducted on shale formations, (at the time of writing this chapter) are single well models. In addition too long, laborious model setup, and computationally expensive simulation process, the interferences between the hydraulic fractures in a multi-lateral pads, makes the history matching for those wells very challenging. Any minor changes in target well's matrix properties and natural/hydraulic fracture characteristics have positive or negative impacts on the history matching results for the offset laterals that should be taken into account.

For performing the history matching, the numerical simulator is run in flowing bottom-hole pressure control mode and the measured daily gas rate productions are set as a target to be matched. Some of key reservoir characteristics (e.g. matrix porosity, natural fracture porosity, natural fracture permeability, rock fraction etc.), hydraulic fracture properties (e.g. HF length and conductivity) and some other completion related parameters such as skin factor are tuned manually to achieve a satisfactory history match for all individual laterals and the entire pad.

Figure 9.4 shows the history matching results for the six horizontal laterals in the WVU Pad. In all plots, dots refer to measured daily gas production (Mscf/day) and the solid lines show the simulation result.

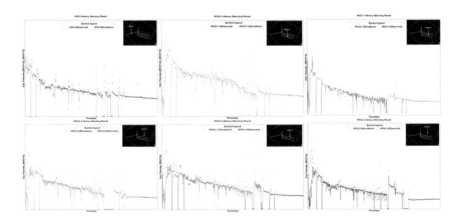

Fig. 9.4 History matching result for all 6 laterals in the WVU pad

9.3 Smart Proxy Modeling

Smart Proxy Modeling is an innovative application of data-driven modeling technology in numerical simulation and modeling. It is a complete paradigm shift on how proxy models are developed in order to maximize the utility of numerical simulation models. In this chapter we demonstrate how a Smart Proxy can be developed for the numerical reservoir simulation model generated for the shale gas wells presented here.

9.3.1 A Short Introduction to Smart Proxy

Proxy modeling is one of the most widely used methods to replicate the functionalities of high fidelity numerical simulation models and to assist in the master development planning, quantification of uncertainties, operational design optimization, and history matching. Most frequently used proxy models in oil and gas industry are reduced order models and response surfaces that reduces simulation run time by approximating the problem and/or the solution space [83]. Response surfaces that are essentially a statistic-based proxy model require hundreds of simulation runs that fits numerical simulation outputs to predetermined functional forms, in order to be used for uncertainty analysis and optimization purposes.

As stated by Mohaghegh [84], there are two well-known problems that are associated with statistics, especially when it is applied to problems with well-defined physics behind it: (a) the issue of "correlation versus causality", (b) imposing a predefined functional form such as linear, polynomial, exponential, etc. to the data that is being analyzed. This approach will fail when data representing the nature of a given complex problem does not lend itself to a predetermined functional form and/or it changes behavior several times.

Smart proxy takes a different approach to building proxy models. In this approach, unlike reduced order models, the physics is not reduced and the space–time resolution is not compromised and instead of using predefined functional forms that are frequently used to develop response surfaces, a series of machine learning algorithms that conform to the system theory are used for training with ultimate goal of accurately impersonating the intricacies and nuances of a developed shale numerical reservoir simulation model.

Ensemble of multiple, interconnected adaptive neuro-fuzzy systems create the core for the development of these models. Both artificial neural networks and Fuzzy Systems were covered in the earlier chapters in this book. Neuro-fuzzy systems are combination and integration of these two techniques. Recently, several papers have been published on the development and validation of Smart Proxies [84–89].

9.3.2 Cluster Level Proxy Modeling

Inclusive spatiotemporal database generation is the starting point and the most important step toward building a smart proxy model. A comprehensive spatiotemporal dataset is generated by coupling the matrix, natural fracture properties and hydraulic fracture characteristics with desorption features as well as operational constraints, which are used in the numerical simulator as input data, with the calculated gas production rate from the simulation model. This database includes details of fluid flow in shale system that the smart proxy model needs to learn. Once the developed proxy model is validated using blind simulation runs, it can be used for reservoir management and planning purposes.

In order to make a comprehensive and more realistic database, in addition to the history-matched model, nine additional realizations are defined to fully capture the uncertainty domain for the study area that includes 77 horizontal wells (Fig. 9.3). The generated spatiotemporal database is used for teaching highly nonlinear and complex shale system behavior to the multilayer feed-forward back-propagation neural networks. The fully calibrated neural network is able to regenerate the numerical simulation output (production profiles) for all 169 clusters of hydraulic fractures for six horizontal laterals that are included in the numerical simulation model.

The shale gas production behavior (i.e. changes in flow regime) as a function of time, is presented in form of three data driven models at hydraulic fracture cluster level with different time resolutions (daily for the first two months, monthly for the first 5 years and yearly for 100 years of production). Figure 9.5 shows the detailed workflow for the data driven models development. Smart Proxy is a combination of multiple data driven models.

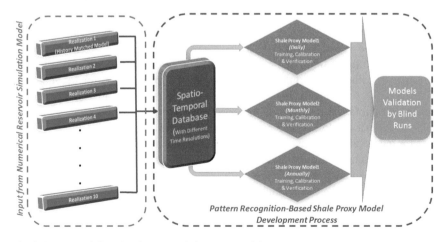

Fig. 9.5 The workflow for the smart shale proxy model

Table 9.1 Main input data used for the smart (data-driven) shale proxy models development

Matrix porosity [0.054–0.135]	Matrix permeability [0.0001–0.00097 (md)]	Natural fracture porosity [0.01–0.04]	Natural fracture permeability [0.001–0.01 (md)]	Sigma factor [0.005–0.6]
Hydraulic fracture height [100–125 ft]	Hydraulic fracture length [200–1100 ft]	Hydraulic fracture conductivity [0.5–5.4 (md-ft)]	Rock density [100–180 (lb/ft^3)]	Net to gross ratio [0.75–0.98]
Longmuir volume [40–85 (scf/ton)]	Longmuir pressure [600–870 psi]	Diffusion coefficient [0.5–2.8 (ft^2/day)]	Sorption time [1–250 (day)]	Initial reservoir pressure [3000–4288 (psi)]

A complete list of inputs and corresponding ranges obtained from the histogram of distributed actual properties for 77 wells that are used for spatiotemporal dataset development is shown in Table 9.1. In order to consider the possible changes in operational constraints and include them in the database, ten different bottom-hole pressure profiles are designed with constant (200, 250, 300, 350 and 400 psi), increasing (from 150 to 550 psi) and declining (from 2500 to 90 psi) trends.

In each simulation run, each time step with corresponding static and dynamic information generates a unique case for the training and validation purposes. In order to take into account the impact of different grid block's properties on each cluster production a "Tier system" is defined. Three different tiers are defined and a property for each tier is calculated by upscaling the properties of all the grid blocks in the corresponding tier. Figure 9.6 (left) illustrates the Tier systems and shows three types of tiers that are used in this study.

- Tier 1: Includes all the refined grid blocks for a target hydraulic fracture cluster.
- Tier 2 (− and +): Covers the rest of grid blocks that are extended to the north (+) and south (−) of the Tier 1, all the way toward the reservoir boundary or a hydraulic fracture from an offset lateral, both laterally and vertically.
- Tier 3 (− and +): Covers the rest of grid blocks that are not covered by Tiers 1 and 2.

Additionally, the interference effect between the clusters can be captured by dividing them into four different classes based on their relative location. The following is the definition of each cluster type and the schematic is shown in Fig. 9.6 (right).

- Type 1: This type of cluster has only one offset cluster that shares part of its drainage region.
- Type 2: The second type of cluster has two offset clusters causes the drainage area to be shared more than the Type 1.
- Type 3: Three neighboring clusters bound the third type thus; the drainage area will be shared more than Type 1 and 2.
- Type 4: Four neighboring clusters bound the last type and drainage area will be shared more than the other types.

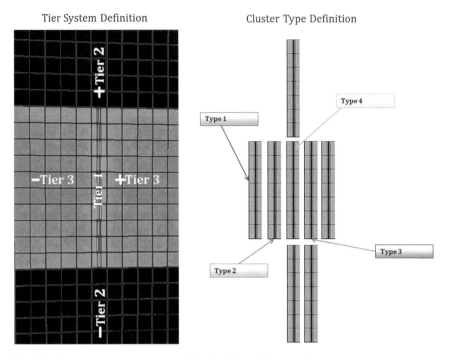

Fig. 9.6 Tier system and cluster type definition (from *left* to *right*)

9.3.3   Model Development (Training and Calibration)

Three data driven models at the hydraulic fracture cluster level with different time resolutions are developed. The first data driven model is designed to regenerate the simulation output for each cluster as well as the entire lateral for the first two months of production with daily time steps (Mscf/day). The second data driven model is developed to mimic the simulation results of methane production for the first 5 years of production on a monthly basis (Mscf/month). Finally, the last data driven model is developed to predict the yearly gas production for 100 years (Mscf/year). Combination of these three data driven models generates the final shale smart proxy model. Following is the details for each data driven model and the results.

9.3.3.1   Smart Proxy Model for Early Time (Transient)

1,690 unique (169 clusters × 10 runs) production profiles with different reservoir properties, sorption attributes (sorption time and Langmuir isotherms), hydraulic fracture component, and operational constraints are generated to create a representative database with 98,020 pairs of input–output data. This dataset is used for training, calibration and validation of a multilayer feed-forward back-propagation

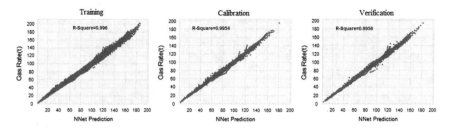

Fig. 9.7 From *left* to *right*, results of training, calibration and validation of the neural network—daily proxy model

neural network with 55 hidden neurons to perform pattern recognition in nonlinear and multi-dimensional problem.

During the training, calibration, and validation process, the most prominent input parameters for the daily gas production from shale are identified. According to this analysis, the hydraulic fracture conductivity and natural fracture permeability are among the most influential parameters for the first two months of production in the Marcellus shale. Moreover, as it is expected, the sorption time and the Langmuir pressure and Langmuir volume which, control desorption, diffusion, and adsorbed gas content of shale, have the minimal impact on the first two months of production.

The spatiotemporal database is partitioned to a 70 % training 15 % calibration and 15 % verification. Figure 9.7 shows the cross plots for proxy model and the numerical simulation output values of daily gas production rate (Mscf/day) for training, calibration and verification steps (from left to right). In these plots, x axis corresponds to the neural network predicted gas rate and the y axis shows the gas rate generated by the reservoir simulator.

The result with an R^2 of more than 0.99 in all steps shows the successful development of daily basis shale proxy model. Figure 9.8 show some examples of the comparison of reservoir simulation output for daily gas production rate (Mscf/day) and those generated by the Smart Proxy model for some of the clusters while Fig. 9.9 shows similar results plotted for several wells (summing all the results for all the clusters included in a well).

In all the plots blue dots represent the daily gas rate generated by the numerical simulator and the solid red line is the result from the smart (data-driven) shale proxy model. The results are self-descriptive enough to show the capability of developed proxy model in predicting the daily gas production profile for each hydraulic fracture cluster as well as lateral for the first two months of production. Moreover, the predicted production rate at cluster level can be accounted as synthetic PLT log (Production Logging Tool) at each time step to show the contribution of each cluster on flow.

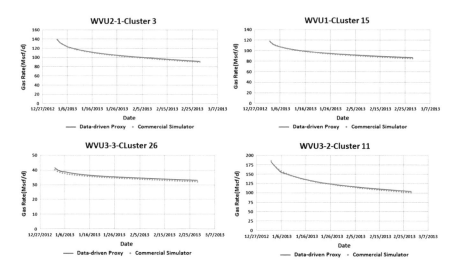

Fig. 9.8 Daily gas production comparison—results from the simulator and shale proxy model for some of the clusters

9.3.3.2 Smart Proxy Model for Middle Time (Late Transient)

Second smart (data-driven) proxy model is developed to regenerate the monthly gas production (Mscf/m) for the first 5 years of the production using the training database with 101,400 pairs of input–output data. The same procedure for building the first proxy model (daily basis) is followed here. In order to understand the effect of each parameter on monthly gas rate (for the first 5 years), Key Performance Indicator (KPI) analysis is performed. According to this analysis, production time, natural fracture permeability, flowing bottom-hole pressure as well as porosity are among the highest ranked parameters during 5 years of production.

It is interesting to see the hydraulic fracture cluster location, which accounts for the interference between the clusters and also sigma factor that controls matrix to fracture flow as the second most important parameters. Alternatively, very low ranking of Langmuir volume and Langmuir pressure as well as diffusion coefficient and sorption time confirm the insignificant impact of sorption features on the production performance for the first 5 years of production.

The cross plots of predicted (by the Smart Proxy) and simulated (by the numerical simulator) values of the monthly gas flow rate (Mscf/month) for training, calibration, and verification steps (from left to right) are shown in Fig. 9.10. In these plots, x axis corresponds to the monthly gas production rates generated by the smart proxy model and the y axis shows the simulated gas rate by the numerical simulator.

In all the steps (training, calibration and verifications), R^2 of more than 0.99 shows the successful development of monthly shale smart proxy model. Figure 9.11 show some examples of the comparison of reservoir simulation output

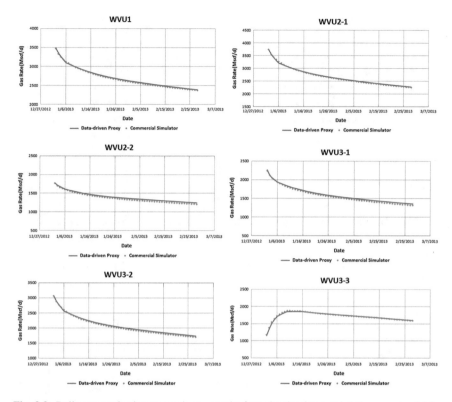

Fig. 9.9 Daily gas production comparison—results from the simulator and shale proxy model for some of the laterals

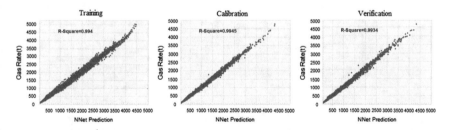

Fig. 9.10 From *left* to *right*, results of training, calibration and validation of the neural network—monthly proxy model

for monthly gas production rate (Mscf/m) and those generated by the Smart Proxy model for some of the clusters while Fig. 9.12 shows similar results plotted for several wells (summing all the results for all the clusters included in a well). In all the plots, blue dots represent the annual gas rate generated by the numerical simulation model, while the solid red line demonstrates the proxy model result.

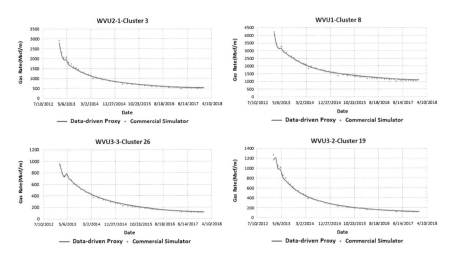

Fig. 9.11 Monthly gas production comparison—results from the simulator and shale proxy model for some of the clusters

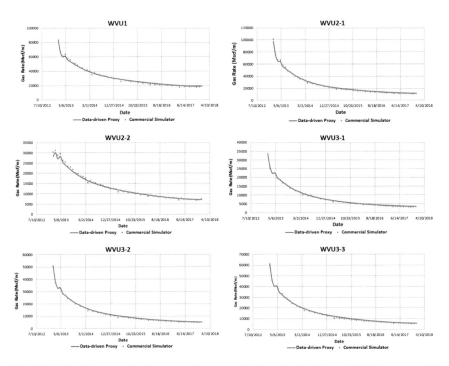

Fig. 9.12 Monthly gas production comparison—results from the simulator and shale proxy model for some of the laterals

9.3.3.3 Smart Proxy Model for Late Time (Pseudo-Steady State)

The descriptive spatiotemporal database with 169,000 pairs of input–output data is used to develop the third smart shale proxy model which is capable of predicting the annual gas rate production (Mscf/year) for 100 years. Production time, sorption time, Langmuir isotherm, natural fracture and matrix permeability and NTG are ranked as the most influential parameters on 100 years production performance of shale gas well.

Figure 9.13 illustrates the cross plots comparing simulated (by numerical simulator) and those generate by the smart proxy for the annual gas production rate (Mscf/year) for training, calibration, and verification steps (from left to right).

In these plots, x axis corresponds to the annual gas production generated by the smart proxy and the y axis shows the simulated annual gas rates by the numerical simulator. The calculated R^2 for training, calibration, and verification results are around 0.99.

Figure 9.14 show some examples of the comparison of reservoir simulation output for annual gas production rate (Mscf/year) and those generated by the Smart Proxy model for some of the clusters while Fig. 9.15 shows similar results plotted for several wells (summing all the results for all the clusters included in a well). In all the plots, blue dots represent the annual gas rate generated by the numerical simulation model, while the solid red line demonstrates the proxy model result.

As illustrated in Figs. 9.14 and 9.15, smart proxy model is successfully regenerated yearly gas rate production profile (red solid line) estimated by a commercial numerical simulator (blue dots) for the 100 years of production at cluster and lateral level.

The only problem that can be observed in almost all the cases is that the smart proxy model for yearly production could not capture the transient behavior that is happening during the first 5 years of production. This was the reason behind developing the first two model. In order to address this problem, monthly and annual based shale proxies are combined. In other word, the first 5 years of production in annual proxy is replaced by the corresponding values that are already generated by monthly proxy model. In this case, the results are improved significantly.

Figures 9.16 and 9.17 show this improvement for four clusters as an example (Fig. 9.14) and the whole WVU2-2 lateral as well (Fig. 9.15).

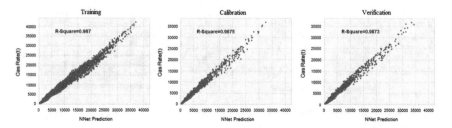

Fig. 9.13 From *left* to *right*, results of training, calibration and validation of the neural network—monthly proxy model

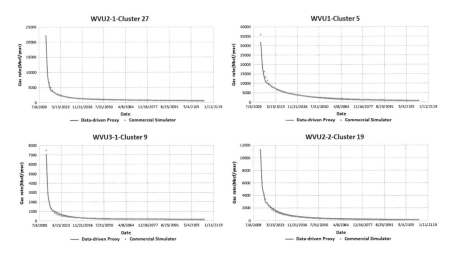

Fig. 9.14 Annual gas production comparison—results from the simulator and shale proxy model for some of the clusters

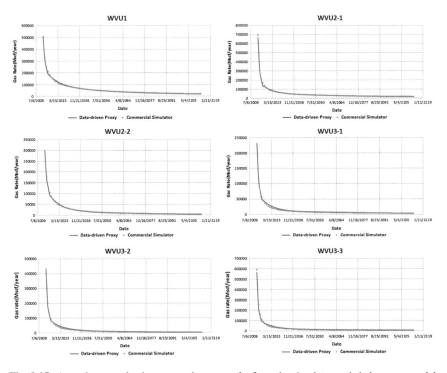

Fig. 9.15 Annual gas production comparison—results from the simulator and shale proxy model for some of the laterals

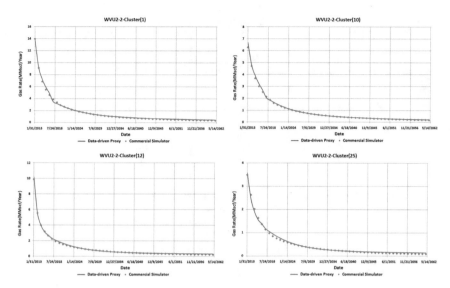

Fig. 9.16 Combined monthly and yearly proxies—yearly gas rate production for four clusters located in WVU-2-2 lateral

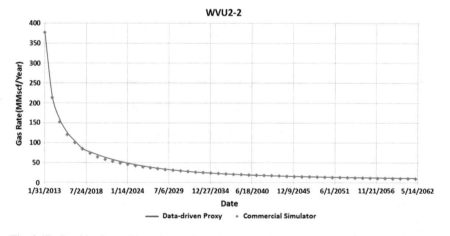

Fig. 9.17 Combined monthly and annual proxies—annual gas rate production WVU-2-2 lateral

9.3.4 Model Validation (Blind Runs)

During smart proxy model development, some of the data were not included in the training set and used for calibration and verification purpose. Nevertheless, in order to perform additional test to examine the predictive capability of the developed smart proxy models, a new simulation run within the uncertainty domain is

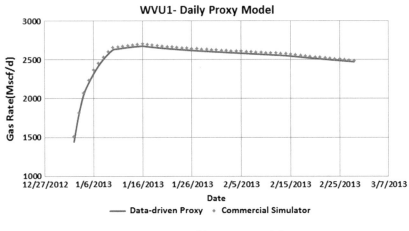

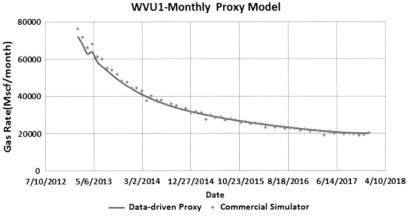

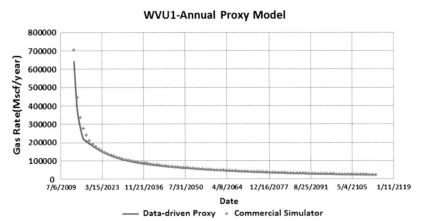

Fig. 9.18 Data-driven proxy models validation using blind run for WVU1 for different time resolutions of daily, monthly and annually (from *top* to *bottom*)

designed that is completely different from the previous runs, which were used during proxy model development process (Training, calibration and primary validation). This is called a "Blind Run".

Figure 9.18 shows the predictive capability of the developed smart proxy models when applies to a blind run. In this figure the simulated gas production rates (blue dots) for different time resolutions of daily, monthly and yearly (from top to bottom) are compared with the smart proxy models (red lines).

Well WVU1 was chosen for demonstration, but similar results were achieved for all the wells. Good results confirm the robustness of developed smart proxy model that it can accurately mimic the behavior of numerical simulation model. This smart proxy can be executed thousands of times in seconds to generate thousands of production profiles with different time resolutions for detailed and quick uncertainty assessment. Performing such a task is completely impractical for the numerical reservoir simulation, since s ingle run of the numerical simulator takes tens of hours.

Chapter 10
Shale Full Field Reservoir Modeling

A quick look at the history of reservoir simulation and modeling indicates that developing full field models (where all the wells in the asset are modeled together as one comprehensive entity) is the common practice for almost all prolific assets. There are many reasons that full field models are developed for prolific assets. Reasons for developing full field models include using the maximum static (geo-logic, geophysics, and petro physics) information available to build the underlying high resolution geological model as well as capturing the interaction between wells.

Looking at the numerical reservoir simulation modeling efforts concentrated on shale assets, one cannot help but to notice that almost all of the published studies are concentrated on analyzing production from single wells [29, 77–79, 90, 91]. There are only two published papers that discuss larger number of wells. One includes modeling four wells in an asset [92] and a second one discusses modeling of 15 wells[1] [93]. The question is: "why full field models, or comprehensive reservoir simulation models are not developed for shale?"

The argument to justify the limited approach (single well) to the numerical modeling of shale assets concentrates on two issues, namely computational expense, and lack of interaction between wells due to low permeability of shale. The argument about the computational expense is quite justified. Those that have been involved with numerical modeling of hydrocarbon production from shale can testify that even modeling a single well that on the average includes 45 clusters of hydraulic fractures (15 stages assuming three clusters per stage) can be a nightmare to set up and run. If Explicit Hydraulic Fracturing (EHF) approach for modeling is used, the model can take tens of hours for a single run (assuming hundreds of parallel CPUs are available and are used for computational purposes, since it is next

---

This Chapter is Co-Authored by: Dr. Soodabeh Esmaili, Devon Energy.

---

[1]There might be other publications that have been published after this book. Also, it is possible (although with low probability) that authors have missed published articles that include modeling of large number of well. Authors' search of the available sources did not reveal such publications.

© Springer International Publishing AG 2017
S.D. Mohaghegh, *Shale Analytics*, DOI 10.1007/978-3-319-48753-3_10

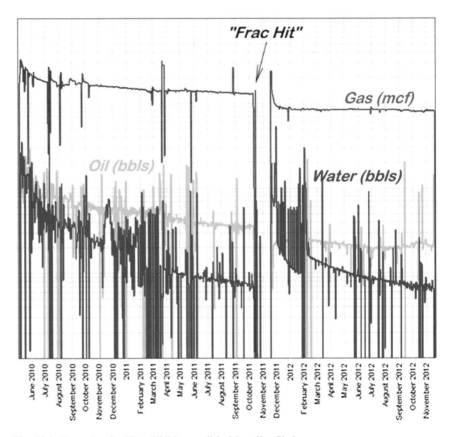

Fig. 10.1 Example of a "Frac Hit" in a well in Marcellus Shale

to impossible to accomplish such a task on a single CPU), therefore building the geological model that would include details of every single cluster of hydraulic fractures (local grid refinement) for an asset with hundreds of laterals is computationally prohibitive. Furthermore, since the nature of shale rock is defined by its very low permeability, minimal interaction between wells is expected and therefore, this logic is used to justify performing single well or sector modeling.

While the first reason (computation expense and manpower required for performing full field modeling) seems to be a legitimate and realistic reason for performing single well (or sector) modeling (specifically for independents, or companies with limited acreage and/or limited engineering resources), the second reason is merely an excuse with limited merit. It is a well-established fact that shale wells do communicate with one another during production. It is shown that the communication takes place between laterals from the same pad as well as the laterals from offset pads. "Frac Hit[2]" is a common occurrence of such interaction. Furthermore, our studies of full field shale assets have clearly shown the importance of including the impact of interference between wells. Figure 10.1 is a "Frac Hit" example from Marcellus Shale.

Therefore, in order to take maximum advantage of the investment that is made and the data that is collected during the field development in a shale asset and to capture interaction between wells and the impact of reservoir discontinuities (faults) on production, it is important to develop full field models for shale assets. Since a comprehensive full field model for a shale asset may require tens (if not hundreds) of millions of grid blocks (for numerical simulation), one may have to look elsewhere for alternative solutions. Data-driven reservoir modeling (Also known as Top-Down Modeling) provides such an alternative.

10.1 Introduction to Data-Driven Reservoir Modeling (Top-Down Modeling)

Top-Down Modeling, a recently developed data-driven reservoir modeling technology [94], (a book named "Data-Driven Reservoir Modeling" is currently in print and will soon be published by the Society of Petroleum Engineers—SPE) is defined as a formalized, comprehensive, multivariant, full field, and empirical reservoir model, which take into account all aspects of production from shale including reservoir characterization, completion, and hydraulic fracturing parameters as well as production characteristics. Despite the common practice in shale modeling using a conventional approach, which is usually done at the well level [95], this technique is capable of performing history matching for all individual wells in addition to full field by taking into account the effect of offset wells.

There are major steps in the development of a Top-Down Model for a shale reservoir that are enumerated as follows:

- **Spatio-temporal database development;** the first step in developing a data-driven shale reservoir model is preparing a representative spatio-temporal database (data acquisition and preprocessing). The extent at which this spatio-temporal database actually represents the fluid flow behavior of the reservoir that is being modeled, determines the potential degree of success in developing a successful model. The nature and class of the data-driven full field reservoir model for shale is determined by the source of this database. The term spatio-temporal defines the essence of this database and is inspired from the physics that controls this phenomenon [85]. An extensive data mining and analysis process should be conducted at this step to fully understand the data that is housed in this database. The data compilation, curation, quality control,

---

[2]A "Frac Hit" is when injected hydraulic fracturing fluid from one well shows up at, and interferes with production from another well.

and preprocessing is one of the most important and time-consuming steps in developing a data-driven full field reservoir model.

- **Simultaneous training and history matching of the reservoir model;** in conventional numerical reservoir simulation the base model will be modified to match production history, while data-driven full field reservoir modeling starts with the static model and try to honor it and not modify it during the history matching process. Instead, we will analyze and quantify the uncertainties associated with this static model at a later stage in the development. The model development and history matching in data-driven full field reservoir model are performed simultaneously during the training process. The main objective is to make sure that the data-driven full field reservoir model learns fluid flow behavior in the shale reservoir being modeled. The spatio-temporal database developed in the previous step is the main source of information for building and history matching the data-driven full field reservoir model.

  In this work, an ensemble of multilayer neural networks is used [44]. These neural networks are appropriate for pattern recognition purposes in case of dealing with nonlinear cases. The neural network consists of one hidden layer with different number of hidden neurons, which have been optimized based on the number of data records and the number of inputs in training, calibration, and verification process, as covered in the previous chapters in this book.

  It is extremely important to have a clear and robust strategy for validating the predictive capability of the data-driven full field reservoir model. The model must be validated using completely blind data that has not been used, in any shape or form, during the development (training and calibration) of the Top-Down Model (TDM). Both training and calibration datasets that are used during the initial training and history matching of the model are considered non-blind. As noted by Mohaghegh [85], some may argue that the calibration—also known as testing dataset—is also blind. This argument has some merits but if used during the development of the data-driven full field reservoir model can compromise validity and predictability of the model and therefore such practices are not recommended.

- **Sensitivity analysis and quantification of uncertainties;** during the model development and history matching that was defined above, the static model is not modified. Lack of such modifications may present a weakness of this technology, knowing the fact that the static model includes inherent uncertainties. To address this, the data-driven full field reservoir model workflow includes a comprehensive set of sensitivity and uncertainty analyses.

  During this step, the developed and history matched model is thoroughly examined against a wide range of changes in reservoir characteristics and/or operational constraints. The changes in pressure or production rate at each well are examined against potential modification of any and all the parameters that have been involved in the modeling process. These sensitivity and uncertainty analyses include single- and combinatorial-parameter sensitivity analyses, quantification of uncertainties using Monte Carlo simulation methods and finally development of type curves. All these analyses can be performed on individual wells, groups of wells or for the entire asset.

- *Deployment of the model in predictive mode;* similar to any other reservoir simulation model, the trained, history matched, and validated data-driven full field reservoir model is deployed in predictive mode in order to be used for reservoir management and decision-making purposes.

10.2 Data from Marcellus Shale

The study presented here focuses on part of Marcellus shale that includes 135 horizontal wells within more than 40 pads which have different landing targets, well lengths and reservoir properties.

10.2.1 Well Construction Data

Since drilling multiple wells from a pad is a common practice in the most shale assets and given the fact that the horizontal wells drilled from a pad experience different interaction with their offsets, three types of laterals have been defined.

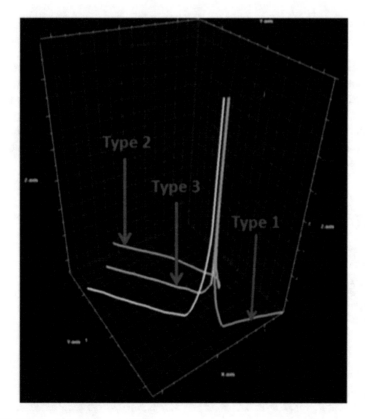

Fig. 10.2 Three well types from a single pad

Figure 10.2 shows the configuration of the three types of wells drilled from a single pad. Based on this definition a new parameter was added to the dataset marked as "Well-Type." This parameter was assigned values such as 1, 2, or 3 in order to incorporate the "Well-Type" information.

Following is a brief description of the three "Well-Types" used in the data-driven full field model.

- *Type One Lateral:* This type of lateral has no neighboring laterals and does not share drainage area (volume). It does not experience any "Frac Hits" from wells in the same pad (it might experience Frac-Hit from a lateral from an offset pad) and its reach will be as far as its hydraulic fractures propagate.
- *Type Two Lateral:* The type two lateral has only one neighboring lateral and therefore; it shares part of the drainage area (volume). Furthermore, "Frac Hits" are possible from laterals in the same pad (it might experience Frac-Hit from a lateral from an offset pad).
- *Type Three Lateral:* The type three lateral is bounded by two neighboring laterals thus; the drainage area (volume) will be shared and "Frac Hits" are possible from both sides in the same pad. If a type three lateral experience a Frac-Hit from an offset pad, it will be from a different depth.

10.2.2   Reservoir Characteristics Data

Marcellus shale in this part of the state of Pennsylvania consists of two prolific layers known as Upper Marcellus (UM) and Lower Marcellus (LM) which are separated by a thin bed limestone layer known as Purcell. Based on the well deviation and completion strategy, one or both layers may be exposed to the production. Reservoir characteristics of each layer including matrix porosity, matrix permeability, pay thickness, net to gross (NTG), initial water saturation, and total organic content (TOC) of each well was given by the operator.

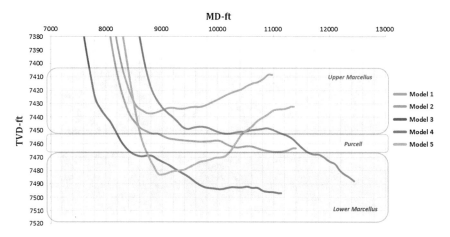

Fig. 10.3 Different well configurations in spatio-temporal database

In order to have a consistent value for each of these properties in well locations, it was assumed that the properties inherited from the completed zone. For instance, if the well is landed and completed in Upper Marcellus, the reservoir properties of this layer should be taken into account. Based on this assumption, five different well configurations Fig. 10.3 were defined and the reservoir characteristics were estimated.

- **Model 1**—these types of wells were landed and completed in Purcell. Because of its low thickness and brittleness, the fracture propagation usually occurs to the upper and lower layers, and therefore, it was assumed that both the UM and LM are contributing into production. The total thickness of UM and LM and the weighted average (Eqs. 10.1 and 10.2) for the rest of the properties were used in these wells. *Calculating the average static property when both Upper and Lower Marcellus formations are completed.*

$$\text{Average Property}(\emptyset, K, \text{TOC}) = \frac{(\text{Property of UM} * h_{\text{UM}}) + (\text{Property of LM} * h_{\text{LM}})}{\text{Total Thickness}}$$

$$(10.1)$$

Calculating the average water saturation when both Upper and Lower Marcellus formations are completed.

$$\text{Average of Water Saturation} = \frac{(S_{\text{W}} \text{ in UM} * h_{\text{UM}} * \emptyset_{\text{UM}}) + (S_{\text{W}} \text{ in LM} * h_{\text{LM}} * \emptyset_{\text{LM}})}{(h_{\text{UM}} * \emptyset_{\text{UM}}) + (h_{\text{LM}} * \emptyset_{\text{LM}})}$$

$$(10.2)$$

- **Model 2**—these types of wells were landed and completed in UM, therefore this is the only layer contributing into production of these wells and the reservoir characteristics of this layer was considered.
- **Model 3**—these well types are very similar to model 2 but they have landed and completed in LM, so only the reservoir characteristics of this layer were taken into account.
- **Model 4**—these wells, as shown in Fig. 10.3, have down-dip trajectories and have been completed in all three layers although the number of stages completed in each layer is different. As a result, the number of stages should be a factor in estimating the average properties. For these wells, the average of reservoir characteristics based on Eq. (10.3) and total thickness of UM and LM was considered. *Calculating the average static property based on number of stages in each formation.*

Avearage Property

$$= \frac{\left[\left(\left(\text{No.of stages completed in UM} + \frac{\text{No.of Stages completed in Purcell}}{2}\right) * \text{Property of UM} * h_{\text{UM}}\right) + \left(\left(\textit{No.of stages complted in LM} + \frac{\text{No.of Stages completed in Purcell}}{2}\right) * \text{Property of LM} * h_{\text{LM}}\right)\right]}{((\text{No. of Stages completed in UM} * h_{\text{UM}}) + (\text{No. of Stages Completed in LM} * h_{\text{LM}}))}$$

$$(10.3)$$

- **Model 5**—these wells have an up-dip deviation and have also been completed in all three layers therefore the same equation as Model 4 was used to estimate the reservoir characteristics.

10.2.3 Completion and Stimulation Data

The completion and stimulation data of the wells include shot density, perforated/stimulated lateral length, number of stages, the amount of injected clean water, rate of injection, injection pressure, amount of injected slurry, etc. Since the production is available on a per well basis, the volumes of fluid and proppant for multiple hydraulic fracture stages performed on the same well were summed while the rates and pressures for these cases were averaged.

The final dataset include more than 1200 hydraulic fracturing stages (approximately 3700 clusters of hydraulic fracturing). Some wells have up to 17 stages of hydraulic fracturing while others have been fractured with as few as four stages. The perforated lateral length ranges from 1400 to 5600 ft. The total injected proppant in these wells ranges from a minimum of about 97,000 lbs. up to a maximum of about 8,500,000 lbs. and total slurry volume of about 40,000–181,000 bbls.

10.2.4 Production Data

The production history of the wells contains the dry gas rate, condensate rate, water rate, casing pressure, and tubing pressure in daily format. The maximum and minimum production history is about 5 years and one and half years, respectively. Because of scattered condensate rates and low condensate to gas ratio (maximum is about 16 STB/MMCF), this data was combined to the dry gas (Eq. 10.4) and the rate of rich gas was estimated for the wells.

Formulation used to combine to the dry gas and the rate of rich gas.

$$GE_{Cond} = 133,800 \frac{\gamma_o}{M_o} \frac{SCF}{STB} \tag{10.4}$$

Where condensate specific gravity and condensate molar density are calculated as follows:

Formulation to calculate condensate specific gravity.

$$\gamma_o = \frac{141.5}{\text{Condensate API} + 131.5} \tag{10.5}$$

Table 10.1 Data available in the dataset that includes location and trajectory, reservoir characteristics, completion, hydraulic fracturing, and production details

Formulation to calculate condensate Molar Density.

$$M_o = \frac{44.43\,\gamma_o}{1.008 - \gamma_o} \tag{10.6}$$

The API degree of condensate was reported by operator as 58.8 at reservoir temperature. In order to remove the associated noises with daily production rates, the monthly basis of this information was used in database. The corresponding monthly wellhead pressure and water rate was also prepared. It has to be noted that, as a common practice in Marcellus, the production is happening from casing in the first couple of months of production and then it is switched through tubing. Therefore, in calculating the average wellhead pressure, this was taken into account. The initial spatio-temporal database after the mentioned calculations, has six main groups as well information, reservoir characteristics, geomechanical properties, completion data, stimulation data, and production and operational constraints. Table 10.1 shows the final list of all data that was used to generate the spatio-temporal database.

10.3   Pre-modeling Data Mining

Pre-modeling data analyses that were performed for Top-Down Modeling is similar to the pre-modeling analyses that are performed during the shale production optimization technology (SPOT). Since this type of data analyses was covered in detail in a previous chapter of this book [8.3 Well Quality Analysis (WQA), 8.4 Fuzzy Pattern Recognition, and 8.5 Key Performance Indicators (KPIs)], they will not be repeated in this chapter.

10.4   TDM Model Development

Similar to other data-driven models, Top-Down Model includes training, calibration, and validation that will be covered in the following sections.

10.4.1   Training and Calibration (History Matching)

During the training and history matching process of the data-driven full field reservoir modeling approach inclusion and exclusion of multiple parameters were examined in order to determine their impact on model behavior. Figure 10.4 includes a flowchart that shows the evolution process of developing the data-driven Marcellus shale full field reservoir model. It starts from the base model (where most of the parameters are included as our first shot) to converge to the best history matched model where optimum number of inputs are identified.

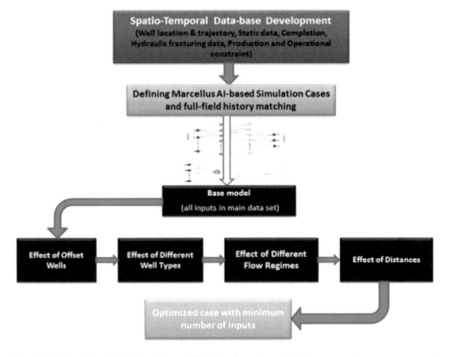

Fig. 10.4 Marcellus shale data-driven full field reservoir model history matching process

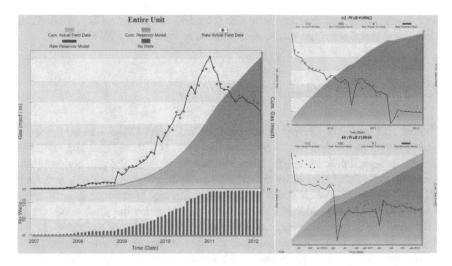

Fig. 10.5 History matching result for entire field (*left*), good (*top right*), and bad (*bottom right*) well

Acceptable history matching result for entire field and for individual wells were achieved and are shown in Fig. 10.5. The plot on the left in Fig. 10.5 represent the full field history matching result while the graphs on the right show examples of good(top) and bad(bottom) result for selected wells. In this figure (all three graphs), the orange dots represent the actual (measured) monthly rates while the green solid line shows the results generated by the Top-Down Model. The orange areal shadow represents the actual (measured) cumulative production (normalized) while the areal green shadow corresponds to cumulative production output (normalized) generated by the top-down model. The red bars (bar chart) at the bottom of the graph on the left show the number of active Marcellus wells as a function of time.

The model is a multilayer neural network that is trained using a back-propagation technique. Data were partitioned into 80 % training dataset, 10 % calibration, and 10 % verification. The cross-plot for top-down model versus actual

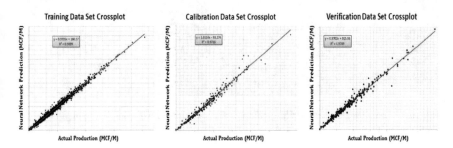

Fig. 10.6 Neural network training, calibration, and verification results-(R^2 of 0.99, 0.97, and 0.975, respectively)

field measurements of monthly flow rate (Mscf/m) are shown in Fig. 10.6. These plots show that the trained model also work very well for the blind data.

10.4.2 Model Validation

Figures 10.5 and 10.6 clearly show that the Top-Down Model can be trained to history match production from a large number wells in a shale field. The validation is done through a process called "Blind History Matching." The production history from this shale field was available from August 2006 to February 2012. In order to validate this history matched model the "Blind History Matching" calls for using part of the historical production to train and history match the production and the tail end of the production is left out to be used as blind history match.

Therefore, the Top-Down Model was trained and history matched August 2006 to October 2011. Then the trained and history matched Top-Down Model is

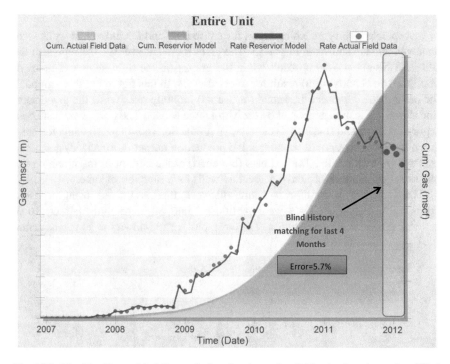

Fig. 10.7 The Top-Down Model's prediction for the entire field—the last 4 months—Blind History Match

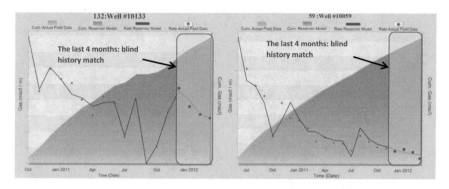

Fig. 10.8 The Top-Down Model's prediction for Well # 10133 and Well # 10059—The last 4 months was blindly history matched

deployed in forecast mode in order to generate production from November 2011 to February 2012. Then the production forecast generated by the Top-Down Model is compared to the historical production from November 2011 to February 2012 (that was originally available but intentionally was left out for validation purposes). Figures 10.7 and 10.8 show the results of the blind history match process that validates the accuracy and robustness of the Top-Down Model developed for this particular shale field.

The above exercise was repeated one more time, this time challenging the Top-Down Model a bit more. In this step, 20 % of the tail end of the production history was removed from the training dataset. Since the length of production for 135 wells varies between 16 and 67 months, therefore, last 4–14 months of production were removed to examine the forecasting ability of the model (Blind history match). Additionally, the Top-Down Model was asked to forecast additional 12 months.

For blind history matching of last 4–14 months, the number of days of production for that period was included in the training set. The averaged flowing wellhead pressure for the last 3 months was used as a constraint for the blind history matching forecasting period (4–14 months and an additional year). Figure 10.9 shows the blind history matching and additional 1 year forecasting results for two wells with 27 and 36 months of production history correspondingly as an example. In this graph, the orange dots represent the actual monthly rate (normalized) while the green solid line shows the Top-Down Model results. The black dots show the actual production data that was removed from the training and tried to be predicted by model. Last 6 and 8 months of production were removed from training (20 % of total month of production) and the TDM could predict the production behavior of those periods with acceptable accuracy. The last 4 months

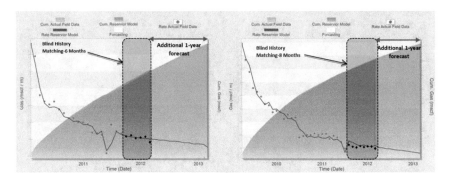

Fig. 10.9 Blind history matching and additional forecasting for two wells with 27 and 36 months of production history, respectively (left to right)

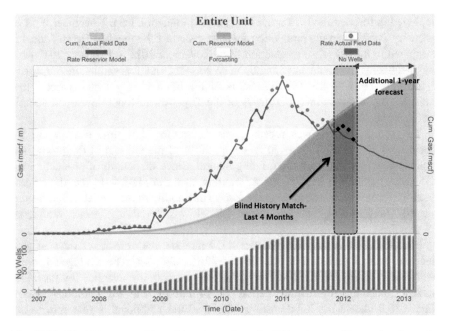

Fig. 10.10 Blind history matching of last 4 months and additional year of forecasting for entire field

of blind data was common between all the wells in the field. Therefore, to show the TDM's predictive capability at the field level, production data (actual and TDM predictions) from all the wells in the field were combined and plotted in Fig. 10.10. The last 4 months data in this figure is blind for all the wells in the field.

Taking validation one-step further, the production performance of a recently drilled well, which was completely blind to the model (was not included during training and initial validation), was predicted and compared with actual field

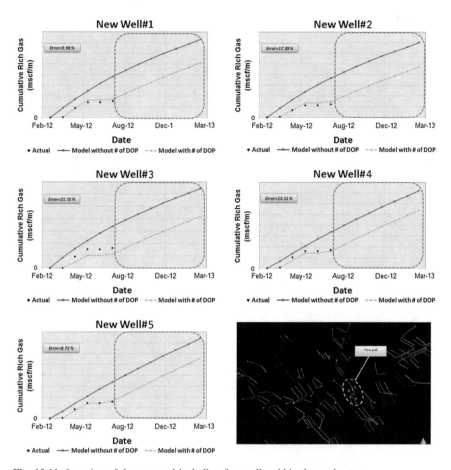

Fig. 10.11 Location of the new pad including five wells within the study area

measurement. Figure 10.11 shows the blind history matching results as well as forecasting for additional year.

By looking at the actual production for the last 4 months, a sudden increase in rate at second month can be clearly observed that might be because of high demand of natural gas over the winter, and then the production followed its natural declining behavior at fourth month. Therefore, the model could predict total production rate for the first and fourth month good enough but it underestimate the total rate for second and third months. The error for predicting the production rate of those 4 months is varying from 1.4 to 9.2 %, for each individual, which shows the capability of model in prediction mode.

Chapter 11
Restimulation (Re-frac) of Shale Wells

Reference publications about refracturing treatments (re-stimulation) before 1990s are sparse. The first published work on re-frac dates back to 1960 [96], followed by another publication in 1973 [97]. Application of data-driven analytics[1] to the subject of hydraulic fracturing in general and re-frac, specifically, originated at West Virginia University in mid-1990s [98–103] and continued into mid-2000s [104–109]. Gas Research Institute[2] started a new re-frac candidate selection project in 1998 that breathed new life into the re-frac technology. Results of this project was extensively published and inspired many new activities in this area [110–113].

Since production of hydrocarbon from shale is intimately tied to hydraulic fracturing it was inevitable that refracturing become a subject of interest for production enhancement in shale. It is a widely accepted notion that at least 40 % of the hydraulic fracture stages in shale wells do not contribute to production. This has been mentioned in articles and interviews with service company representative [114] and has been indicated through microseismic (Fig. 2.12 or similar figures in the literature indicate minor or even lack of seismic activates in some of the stages). Even in the stages that have been properly fracked and are contributing to production, the depletion of the reservoir from hydrocarbon modifies the stresses that controlled the original orientation of the hydraulic fractures and if new frac jobs are pumped into these same stages, chances are that the propagation of new fracs will take different path compared to the original fracs, and new reservoir will be contacted and result in higher production. Figure 11.1 is a schematic diagram of the frac reorientation due to stress modification in the reservoir after partial depletion of the reservoir.

As it was discussed in the previous section of this book, the author (inline with most of the completion expert in the industry today) does not believe in the development of well-behaved, penny-shaped, hydraulic fractures in shale, nevertheless, the general idea that partial depletion will modify the original stresses in the

---

[1]At the time the data-driven analytics was not the popular reference to these series of technologies and they were referred to as artificial neural networks, or intelligent systems.
[2]Now: Gas Technology Institute.

© Springer International Publishing AG 2017

S.D. Mohaghegh, *Shale Analytics*, DOI 10.1007/978-3-319-48753-3_11

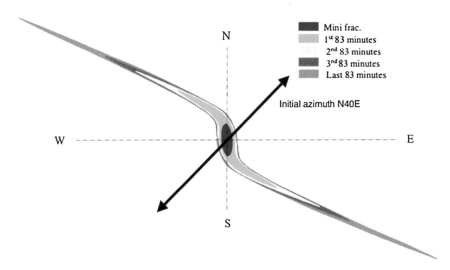

Fig. 11.1 Frac reorientation due to previous production and modification of the stress around the wellbore [115]

rock and provides a new stress filed around the wellbore causing a different propagation path for the new hydraulic fracture, seems to be a sound argument.

Once we accept the idea that re-frac can provide major potential for more production from shale wells by tapping into new reserves (whether it be by fracking stages that were not productive in the first pass, or by propagating into new natural fractures resulting from modification of the stress field) two new questions surface that need to be answered accurately in order to have a successful re-frac program:

1. Given that most shale asset include large number of wells, how would one screen them in order to identify the best re-frac candidates? It should be obvious that not all wells would respond in the same manner to a re-frac.
2. Once the re-frac candidate selection is completed, how would one design a re-frac treatment for a well that has been hydraulically fractured before?

The objective of this chapter of the book is to address the above two questions. We use Shale Analytics and the developments that have been demonstrated in the previous several chapters in this book in order to address the above two questions.

11.1 Re-frac Candidate Selection

The data-driven analytics approach to re-frac candidate selection in shale wells (Shale Analytics) is intimately related to two other topics that were covered in the previous chapters of this book, namely, Sects. 8.6 and 8.7. Upon the development (training and calibration) and validation of a data-driven predictive model, we are in possession of a tool that can predict production (in this case production indicators) from a shale well that is conditioned to a large number of field measurements with little to no human biases involved.

The data-driven predictive model calculates well production that is conditioned to well construction, reservoir characteristics, completion and stimulation (hydraulic fracture) parameters, and operational constraints. This is the only technology that can accurately predict shale well productivity while making the most use of all the field measurements, and by avoiding any and all types of interpretations and use of soft data. Using the data-driven predictive model provides the luxury of being able to modify some of the input parameters and examine their impact on model output (shale well's productivity). This is a powerful tool for understanding the role of different parameters or group of parameters on shale well's productivity. Furthermore, given the fact that the data-driven predictive model has small computational footprint (only a fraction of a second per execution) it can be executed a large number of times in a span of only a few seconds and therefore, provide the means for studies that require large number of model execution such as uncertainty analysis.

For example, if for a given well, the operational constraints are modified, we can learn how the shale well will respond to changes to surface facilities constraints (translated to well-head pressure) and choke sizes. By modifying reservoir characteristics and monitoring the model's response (production from the shale well), we can learn the impact of different reservoir parameters and their role in controlling production from shale wells. Similarly, as was shown in Sect. 8.7 we can assess the importance of different completion practices on the production from a given well. It can easily be seen that the data-driven predictive model is a powerful tool in learning so much about the production behavior of a shale well.

Our philosophy behind the identification of best re-frac candidate wells has to do with "unfulfilled potentials". In other words, in the process of identifying the best candidate wells for re-frac, we will look for the wells that based on our validated, data-driven predictive model could have produced a lot more hydrocarbon, given the way the well was originally constructed and the reservoir characteristics. In this section we demonstrate the use of the data-driven predictive model and the insight gained from the Look-Back analysis, in order to identify and rank re-frac candidates.

The process of "Shale Analytics for Re-frac Candidate Selection" (SARCS) is accomplished using the following four steps:

Step 1 Complete the Look-Back Analysis and generate a table that includes all the wells in the asset (that are being analyzed) with their corresponding P10, P50, and P90 values. Furthermore, the table should include the actual well productivity. A sample of such a table generated for a Marcellus shale asset is shown in Fig. 11.2. In this figure, the last column identifies the probability location of the actual production of the well (Px) as compared to P10, P50, and P90 values.

Index	ISI Coded WellName	MSCF			Actual Well Productivity	Output Probability
		P10	P50	P90		
1	Well #0000001	10,008	5,327	2,839	4,816	58
2	Well #0000002	17,320	13,612	9,523	15,582	27
3	Well #0000003	17,535	14,511	10,876	11,520	85
4	Well #0000004	13,699	10,329	7,256	7,685	85
5	Well #0000005	15,568	12,223	9,199	13,629	29
7	Well #0000007	17,485	14,182	9,306	17,859	6
8	Well #0000008	18,025	15,705	12,136	13,117	82
9	Well #0000009	16,478	13,193	9,253	11,004	75
10	Well #0000010	12,675	8,872	5,461	12,027	14
11	Well #0000011	15,233	11,822	8,212	9,459	79
14	Well #0000014	8,855	6,327	4,341	5,786	62
15	Well #0000015	13,339	10,146	6,999	7,999	80
16	Well #0000016	12,672	9,197	6,128	10,580	32
17	Well #0000017	11,043	8,522	6,344	10,660	14
18	Well #0000018	12,685	9,578	7,161	6,772	94
19	Well #0000019	12,381	9,556	7,154	8,880	63
20	Well #0000020	11,423	8,637	6,480	6,437	90
21	Well #0000021	10,438	6,109	3,481	10,599	9
22	Well #0000022	15,753	12,397	8,749	9,719	82
23	Well #0000023	15,000	11,630	8,082	10,326	70
24	Well #0000024	9,440	5,352	2,808	8,893	13
25	Well #0000025	12,465	7,978	4,601	5,530	80
26	Well #0000026	14,589	9,996	5,931	6,257	88
28	Well #0000028	18,394	15,866	12,364	16,139	46
29	Well #0000029	11,185	7,201	3,997	5,642	72
30	Well #0000030	16,457	13,110	9,630	8,471	96
31	Well #0000031	16,239	12,880	9,626	9,757	89
32	Well #0000032	13,599	9,682	6,779	6,007	95
33	Well #0000033	11,830	7,351	4,226	6,743	57
34	Well #0000034	11,321	6,959	3,820	11,807	8
35	Well #0000035	11,067	6,748	3,730	14,358	1
36	Well #0000036	11,103	7,016	3,809	8,202	35
37	Well #0000037	11,095	6,966	3,998	4,498	85
38	Well #0000038	15,288	12,580	9,038	14,278	23
39	Well #0000039	15,953	11,501	6,715	15,374	15
40	Well #0000040	10,587	7,381	4,676	8,839	28
41	Well #0000041	13,394	9,668	6,308	9,686	50
42	Well #0000042	15,501	12,076	8,381	4,622	100
43	Well #0000043	13,759	9,390	5,570	8,724	58
44	Well #0000044	10,207	6,101	3,834	7,667	29
46	Well #0000046	17,067	14,375	10,006	14,735	44

Fig. 11.2 List of wells from a Marcellus shale asset. Data-driven predictive model was used to calculate P10, P50, and P90 as part of the Look-Back analysis

Index	ISI Coded WellName	P10	P50	P90	MSCF Actual Well Productivity	Potential to (P50)	Potential to (P10)	Output Probability
1	Well #0000001	10,008	5,327	2,839	4,816	-511	-5,192	58
2	Well #0000002	17,320	13,612	9,523	15,582	1,970	-1,738	27
3	Well #0000003	17,535	14,511	10,876	11,520	-2,991	-6,015	85
4	Well #0000004	13,699	10,329	7,256	7,685	-2,644	-6,014	85
5	Well #0000005	15,568	12,223	9,199	13,629	1,406	-1,939	29
7	Well #0000007	17,485	14,182	9,306	17,859	3,677	374	6
8	Well #0000008	18,025	15,705	12,136	13,117	-2,588	-4,908	82
9	Well #0000009	16,478	13,193	9,253	11,004	-2,189	-5,474	75
10	Well #0000010	12,675	8,872	5,461	12,027	3,155	-648	14
11	Well #0000011	15,233	11,822	8,212	9,459	-2,363	-5,774	79
14	Well #0000014	8,855	6,327	4,341	5,786	-541	-3,069	62
15	Well #0000015	13,339	10,146	6,999	7,999	-2,147	-5,340	80
16	Well #0000016	12,672	9,197	6,128	10,580	1,383	-2,092	32
17	Well #0000017	11,043	8,522	6,344	10,660	2,138	-383	14
18	Well #0000018	12,685	9,578	7,161	6,772	-2,806	-5,913	94
19	Well #0000019	12,381	9,556	7,154	8,880	-676	-3,501	63
20	Well #0000020	11,423	8,637	6,480	6,437	-2,200	-4,986	90

Fig. 11.3 Potential P50 and Potential P10 calculated for all wells

Step 2 Subtract "Actual Well Productivity" from the calculated P50 values. This calculated value is called the "Potential to P50" (Fig. 11.3). This is the difference between the actual well production and the expected average production (P50) that should have (and could have) been easily achieved from this well. This is the first indicator of the amount of hydrocarbon that could have been recovered, but was missed during the original set of hydraulic fractures.

Step 3 Subtract "Actual Well Productivity" from the calculated P10 values. This calculated value is called the "Potential to P10" (Fig. 11.3). This is the difference between the actual well production and the expected best production (P10) that could have been achieved from this well. This is the second indicator of the amount of hydrocarbon that could have been recovered, but was missed during the original set of hydraulic fractures.

Step 4 Sort the table based on the "Potential to P50" (This is the missed potential to achieve average production. This must be easiest to achieve) and "Potential to P10" (This is the missed potential to achieve an excellent frac job and. This is achievable, but not very easily) separately. These calculations and ranking are shown in Fig. 11.4.

Step 5 There are three tables in Fig. 11.4. The right-table in Fig. 11.4 is the final ranking of the Re-Frac Candidates. This table is generated using the ranking in the left- and the middle-tables in Fig. 11.4. The reconciliation of the left- and the middle-tables in Fig. 11.4 is accomplished by giving the ranking of "Potential to P50" twice as much value as the ranking of "Potential to P10".

P50 Ranking	ISI Coded WellName	Potential to (P50)	P10 Ranking	ISI Coded WellName	Potential to (P10)	Combined Ranking Score	Final Ranking	ISI Coded WellName
47	Well #0000001	-511	33	Well #0000001	-5,192	4	1	Well #0000042
92	Well #0000002	1,970	83	Well #0000002	-1,738	5	2	Well #0000066
16	Well #0000003	-2,991	23	Well #0000003	-6,015	11	3	Well #0000079
21	Well #0000004	-2,644	24	Well #0000004	-6,014	12	4	Well #0000119
77	Well #0000005	1,406	78	Well #0000005	-1,939	17	5	Well #0000068
112	Well #0000007	3,677	112	Well #0000007	374	20	6	Well #0000065
22	Well #0000008	-2,588	39	Well #0000008	-4,908	23	7	Well #0000135
27	Well #0000009	-2,189	30	Well #0000009	-5,474	24	8	Well #0000030
108	Well #0000010	3,155	98	Well #0000010	-648	29	9	Well #0000067
25	Well #0000011	-2,363	26	Well #0000011	-5,774	29	10	Well #0000085
46	Well #0000014	-541	59	Well #0000014	-3,069	30	11	Well #0000050
28	Well #0000015	-2,147	32	Well #0000015	-5,340	31	12	Well #0000026
76	Well #0000016	1,383	77	Well #0000016	-2,092	40	13	Well #0000032
97	Well #0000017	2,138	103	Well #0000017	-383	41	14	Well #0000136
19	Well #0000018	-2,806	25	Well #0000018	-5,913	48	15	Well #0000080
43	Well #0000019	-676	53	Well #0000019	-3,501	49	16	Well #0000031
26	Well #0000020	-2,200	38	Well #0000020	-4,986	52	17	Well #0000081
118	Well #0000021	4,490	109	Well #0000021	161	55	18	Well #0000003
20	Well #0000022	-2,678	22	Well #0000022	-6,034	62	19	Well #0000022
34	Well #0000023	-1,304	40	Well #0000023	-4,674	63	20	Well #0000018
110	Well #0000024	3,541	100	Well #0000024	-547	64	21	Well #0000037
24	Well #0000025	-2,448	17	Well #0000025	-6,935	65	22	Well #0000025
12	Well #0000026	-3,739	7	Well #0000026	-8,332	66	23	Well #0000004
56	Well #0000028	273	72	Well #0000028	-2,255	76	24	Well #0000011
32	Well #0000029	-1,559	28	Well #0000029	-5,543	83	25	Well #0000008

Fig. 11.4 Sorting the table of wells based on the highest production missing values of potential to P50 (*left-table*) and potential to P10 (*middle-table*). Final ranking of re-frac candidates (*right-table*)

11.2   Re-frac Design

In the previous section, identification and ranking of re-frac candidates were covered. Once the re-frac candidate wells are identified, it is time to design the most appropriate frac job to be pumped. The objective of the design of frac job is to have a treatment that is as close to optimum as possible. The optimum frac job is defined as a frac job that creates (activates) the largest possible network of highly conductive pathways into the wellbore such that it would maximize the amount of hydrocarbon that can be produced. In other words, the quality of the frac job is judged based on the hydrocarbon production that it triggers and sustains. The frac job needs to do what it does in order to contribute to higher, more sustainable hydrocarbon production.

As suggested by the main theme of this book, i.e., data-driven analytics (Shale Analytics), we will learn from the historical data in order to design a new frac job. The premise that we have been basing all our work, suggests that instead of basing our analysis, modeling and design judgments on our today's understanding of the physics of storage and transport phenomena in shale, we will use the field measurements as the basis of our design and analysis. The technology that has been introduced in this book provides the means for using field measurements as sets of input–output records that will eventually guide our modeling and design of frac jobs in shale wells.

To use data-driven analytics as a design tool, we make maximum use of the data-driven predictive model as the objective function of a search and optimization algorithm. Search and optimization algorithms are developed in order to find the optimum combination of input parameters that results in the best output of a given objective function. In this analyses, well productivity is defined as the result (output) of the objective function (a yardstick to compare effectiveness of frac jobs in combination with reservoir characteristics) of our search and optimization algorithm. The combination of well construction, reservoir characteristics, completion practices, operational constraints, and frac parameters serve as the input parameters. A schematic diagram of the flow chart is provided in Fig. 11.5.

There are many search and optimization algorithms that can be used for the purposes identified above. Staying loyal to the main theme of this book, data-driven analytics (Shale Analytics), we will use an evolutionary optimization technique called genetic algorithms for this purpose. Evolutionary computing paradigms provide a rich and capable environment to solve many search, optimization, and design problems. The larger the space of the possible solutions, the more effective would be the use of these paradigms. Evolutionary computing, in general, and genetic algorithms, specifically, are able to combine the exploration characteristics of an effective search algorithm with a remarkable ability of preserving and exploiting the knowledge acquired during every step of the search as a guide to take the next step. This provides an intelligent approach to more efficiently solve search, optimization, and design problems. The re-frac design that is introduced in this section uses genetic algorithm as the search and optimization technology.

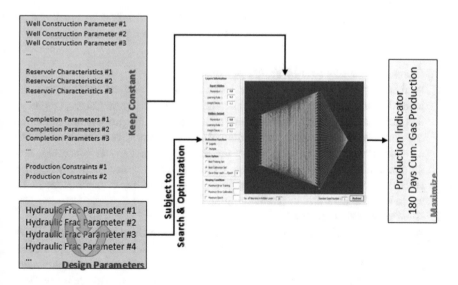

Fig. 11.5 Search and optimization flow chart for the design of optimum re-frac

The optimization of re-frac design that is introduced in this section can be applied to the entire asset, to a group of wells that are identified based on some common theme such as geographic location, reservoir characteristics, or production characteristics (such as common BTU range), or to an individual well. The algorithm that is covered in this section does not change as it is applied to the entire field, group of wells or an individual well. The difference will show up in the population of the parameters (as it will be explained next) during the analyses. For the purposes of explaining the process, we apply the design optimization algorithm to the entire field.

The main distinction between the re-frac design technology proposed here with other methods is that in this technology we let the realities of completion practices in a given asset (field) be the main design driver. Instead of using our (today's) understanding of the completion and hydraulic fracturing in shale in order to design a re-frac treatment, this technology capitalizes on facts collected in the field (field measurements) to drive the treatment design. Since our understanding of interaction between shale's natural fracture network and the induced fractures is quite limited (at best), this technology prevents human bias to enter the design process.

In summary, this technology (for designing re-frac treatment) uses the lessons learned from the past frac jobs in the same asset to generate an optimum design for re-frac treatment. This is done by evolving the new treatment design that is conditioned to all the details (well construction, reservoir characteristics and completion) regarding a given well, through several generations, while keeping track of the characteristics of the most successful individual frac designs (combination of design parameters), then looking for distinctive patterns among these successful individual frac designs.

Following steps are performed for every individual well in order to generate a set of general guidelines for the optimum re-frac treatment design that is applicable to the entire the field

1. *Generating random set of design parameters*

   In this step, the design parameters that are the subject of optimization are identified and separated from other parameters (design or nondesign parameters that are not going to be modified [optimized] during this process are called the constant parameters.). Then, a large number (for example one thousand, 1000) of realization of the combination of the design parameters are generated, randomly. Each realization of the combination of design parameters are coupled with the constant parameters to generate one set of input parameters to be applied to the data-driven predictive model.

2. *Generating model outputs*

   Apply each set of input parameters (total of 1000) to the data-driven predictive model and generate corresponding outputs (production indicators). Upon completion of this step, we have generated 1000 values of the output parameter (production indicator) for each well. Each production indicator represent one randomly generated frac design applied to the well (while keeping all other parameters constant). A schematic diagram of this process is shown in Fig. 11.6.

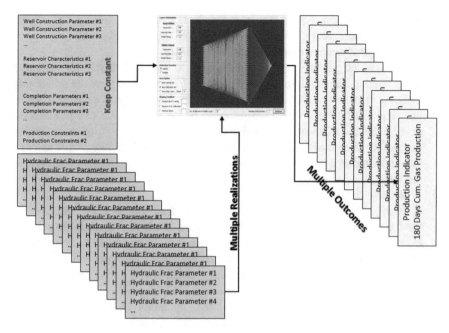

Fig. 11.6 Generating multiple realizations of designs for each individual well

Upon completion of this process for an individual well in the field we have generated 1000 production indicators representing large number of possible outcomes (production) for a single wells with a given set of characteristics (well construction, reservoir characteristics and completion).

3. ***Ranking the model outputs***

The one thousand production indicators (for example, 180 days of cumulative production) are sorted from highest value (production) to lowest with their corresponding combination of design parameters. At this point in time, we have identified the best combination of the design parameters for a given well, from among the 1000 randomly generated designs.

By sorting these combination of design parameters, we now know which set of combinations are responsible for highest production given all the details of this particular well. Remember that by looking at the details of the design parameters, we may or may not understand "why" one set of combination of design parameters are better for this given well, than other combination (better design parameters result in higher production). But it does not matter "why," as long as we know one set is better than the other, we are going to use this knowledge to our advantage.

4. ***Save the best designs***

Upon completion of Step 3, we have completed one generation of evolution. We save the combination of design parameter for the top 10 % of the solutions. By the end of first generation we have now a collection of 100 "good" re-frac designs (combination of design parameters) for this particular well.

5. *Assign reproduction probability based on fitness*

   A reproduction probability value is assigned to each individual solution based on its ranking generated in the previous step. Figure 11.7 shows the probability assignment based on fitness. The values of the probability that are assigned in this step will be used during the selection of the parents that will produce the next generation of the re-frac design solutions. The higher is the reproduction probability of an individual solution, the higher is the probability of that solution to be selected as a parent and play a role in the makeup of the next generation of the individual solutions.

6. *Producing the next generation*

   A large portion of the next generation of the re-frac designs are produced using the top solutions from the previous generation. Each time parents are selected from the previous generation based on their probability values assigned in the last step. Solutions with the higher probability values will have a better chance of being selected as parent to produce the population of the solutions for the new generation. The new generation is produced using genetic operators such as crossover, inversion, and mutation. These operators were explained in the previous sections of this book (Figs. 3.21–3.24). Usually a large portion of the population of the new generation is produced via cross over (about 75 %), where other genetic operators are used to produce a smaller amount of the new

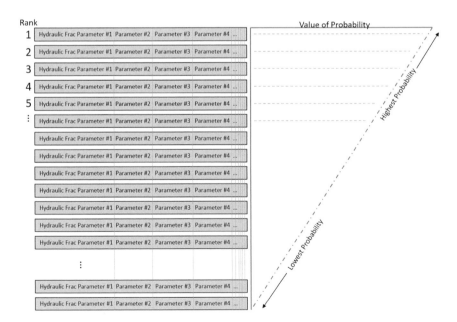

Fig. 11.7 Assigning reproduction probability to each individual solution based on their fitness

generation of solutions (5 % for each inversion and mutation operators). Furthermore, a small percentage of the population of the previous generation is used to move to the next generation, where the value of the probability plays an important role, again.

7. **Repeat from Step 2, until convergence**

Once a new generation of solutions is produced, the algorithm goes back to Step 2 and repeats the process. Each time a cycle of the genetic optimization is completed during the evaluation of the goodness of the solutions that have been generated (Step 2), the algorithm must check to see if convergence has been reached. The algorithm continues only if the convergence has not yet been achieved. Convergence can be defined in multiple ways. For example, convergence can be achieved when no better individual (re-frac design solution) is evolved after several generations.

8. **Analyze the data saved in Step 4**

Let us assume that it takes 145 generations before the optimum re-frac design is evolved using the process enumerated above. Since the top 10 % solutions (re-frac design parameters) of each generation were being saved (100 solutions), and since it took 145 generation to converge to the best solution, upon completion of the above optimization process we have saved 14,500 top solutions. These 14,500 solutions have one thing in common. They all are among the top survivors, as the fittest solutions of their generation. Therefore, they must be doing something right since the definition of a "good solution" in the context of re-frac treatment design is the best combination of hydraulic fracture parameters that results in high production.

Now that the data representing the set of best solutions for a re-frac design is available, we should be able to see if this data contains any specific patterns or trend that can guide us toward the best re-frac design for this specific well. Figure 11.8 is the result of performing the above optimization on an individual shale well producing from Utica shale. This figure includes three of the several parameters that were used in this process. The top-left bar chart in this figure shows the frequency of the "Shot Density" (shots per ft. of completion) in the completion process. This bar chart clearly shows that the overwhelming percent of the top solutions (re-frac designs) have small number of shot density. The top-right bar chart shows a clear pattern of higher values of proppant concentration (lbs/gal/ft. of completion) for the top design solutions for this well, and finally the bottom bar chart shows that the best designs for this well mostly included lower values of injection rates (BPM/ft. of completion).

This figure shows that there is an undeniable pattern in the top re-frac designs that results in high production in this particular well. Therefore, in order to have a successful re-frac, it is recommended that the numbers generated by this optimization algorithm be honored as much as it is operationally feasible. For example for the example presented in Fig. 11.8, it is recommended that the number of shots

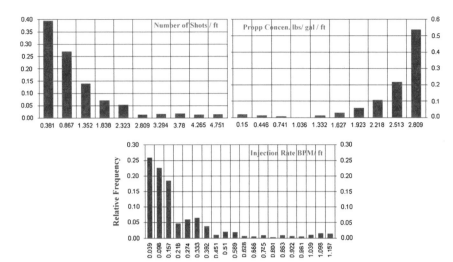

Fig. 11.8 Result of evolving re-frac design solution using a genetic optimization algorithm for a well in Utica shale

per foot of completed lateral length be <1.3, the proppant concentration should be higher than 2.5 lbs per gallon of fluid per foot of completed lateral length and the injection rate be kept at less than 0.15 ballers per minute per foot of completed lateral length.

Bibliography

1. Hunt, G.: How the US Shale Boom Will Change the World. OilPrice.com. (Online) February 15, 2012 (Cited: November 30, 2015). http://oilprice.com/Energy/Natural-Gas/How-the-US-Shale-Boom-Will-Change-the-World.html
2. Mohaghegh, S.D.: Reservoir modeling of shale formations. J. Nat. Gas Sci. Eng. **12**, 22–33 (2013) (Elsevier, s.l.)
3. Mohaghegh, S.D.: A critical view of current state of reservoir modeling of shale assets. In: Society of Petroelum Engineers – SPE, Pittsburgh, PA, 165713 (2013)
4. Kuhn, T.S.: The Structure of Scientific Revolutions. University of Chicago Press, Chicago, IL (1996). 0226458075
5. Kuhn, T.S.: The Essential Tension: Selected Studies in Scientific Tradition and Change. The University of Chicago Press, Chicago, IL (1984). 0-226-45806-7
6. Boulis, A.S., Ilk, D., Blasingame, T.A.: A new series of rate decline relations based on the diagnosis of rate-time data. In: Canadian International Petroleum Conference (CIPC). s.n., Calgary, Alberta (2009)
7. Genesis. The North Texas Barnett Shale Opens New Energy Era, North American Shale Revolution. Energyintel.com. (Online) Sept–Dec 2011. www.energyintel.com
8. Doe, T., Dershowitz, W.: Modeling Complexities Of Natural Fracturing Key In Gas Shales. Oil and Gas Reporter, s.l., Aug 2011
9. Tran, N.H., Rahman, M.K., Rahman S.S.: A Nested Neuro-Fractal0Stochastic Technique for Modeling Naturally Fractures Reservoirs. Society of Petroleum Engineers (SPE), Melbourne, Australia, Oct 2002. SPE 7787
10. Akbarnejad-Nesheli, B., Valko, P., Lee, J.: Relating fracture network characteristics to shale gas reserve estimation. In: Americas Unconventional Resources Conference. Society of Petroleum Engineers (SPE), Pittsburgh, PA June 2012. SPE 154841
11. Li, Y., Wei, C., Qin, G., Li, M., Luo, K.: Numerical simulation of hydraulically induced fracture network propagation in shale formation. In: International Petroleum Technology Conference. Society of Petroleum Engineers (SPE), Bejing, China, Mar 2013. IPTC 16981
12. Weng, X., Kresse, O., Cohen, C., Wu, R., Gu, H.: Modeling of hydraulic fracture network propagation in a naturally fractured formation. In: SPE Hydraulic Fracturing Technology Conference & Exhibition. Society of Petroleum Engineers (SPE), The Woodlands, Texas, Jan 2011. SPE 140253
13. Cheng, Y., Lee, W.J., McVay, D.A.: A New Approach for Reliable Estimation of Hydraulic Fracture Properties Using Elliptical Flow Data in Tight Gas Wells. SPE Reservoir Evaluation & Engineering, Apr 2009. SPE 105767
14. Mattar, L., Gault, B., Morad, K., Clarkson, C.R., Freeman, C.M., Ilk, D., Blasingame, T.A.: Production analysis and forecasting of shale gas reservoirs: case history-based approach. In: SPE Shale Gas Production Conference. Society of Petroleum Engineers (SPE), Fort Worth, Texas, Nov 2008. SPE 119897

© Springer International Publishing AG 2017

S.D. Mohaghegh, *Shale Analytics*, DOI 10.1007/978-3-319-48753-3

15. Johnson, N.L., Currie, S.M., Ilk, D., Blasingame, T.A.: A simple methodology for direct estimation of gas-in-place and reserves using rate-time data. In: SPE Rocky Mountain Technology Conference. Society of Petroleum Engineers (SPE), Denver, Colorado, Apr 2009. SPE 123298

16. Can, B., Kabir, C.S.: Probabilistic Performance Forecasting for Unconventional Reservoirs with Stretched-Exponential Model. Society of Petroleum Engineers (SPE), s.l., Feb 2012. SPE Reservoir Eval. Eng J. SPE 143666

17. Ikewun, P., Ahmadi, M.: Production optimization and forecasting of shale gas wells using simulation models and decline curve analysis. In: SPE Western Regional Conference. Society of Petroleum Engineers (SPE), Bakersfield, California, Mar 2012. SPE 153914

18. Ilk, D., Rushing, J.A., Perego, A.D., Blasingame, T.A.: Exponential vs. Hyperbolic Decline in Tight Gas Sands: Understanding the Origin and Implications for Reserve Estimates Using Arp's; Decline Curves. Society of Petroleum Engineers, s.l. (2008)

19. Valko, P.P., Lee, W.J.: A Better Way To Forecast Production From Unconventional Gas Wells. Society of Petroleum Engineers, s.l. (2010)

20. Duong, A.N.: An Unconventional Rate Decline Approach for Tight and Fracture-Dominated Gas Wells. Society of Petroleum Engineers, s.l. (2010)

21. Mohaghegh, S.D.: Formation vs. Completion: Determining the main drivers behind production from shale? A case study using data-driven analytics. In: Unconventional Resources Technology Conference. Society of Petroleum Engineers (SPE), San Antonio, Texas, July 2015. URTeC 2147904

22. Ilk, D., Rushing, J.A., Blasingame, T.A.: Integration of production analysis and rate-time analysis via parametric correlations—theoretical considerations and practical applications. In: PE Hydraulic Fracturing Conference. Society of Petroleum Engineers (SPE), The Woodlands, Texas, Jan 2011. SPE 140556

23. Al-Ahmadi, H.A., Almarzooq, A.M., Wattenbarger, R.A.: Application of linear flow analysis to shale gas wells—field cases. In: SPE Unconventional Gas Conference. Society of Petroleum Engineers (SPE), Pittsburgh, PA, Feb 2010. SPE 130370

24. Anderson, D.M., Nobakht, M., Moghadam, S., Mattar, L.: Analysis of production data from fractured shale gas wells. In: SPE Unconventional Gas Conference. Society of Petroleum Engineers (SPE), Pittsburgh, PA, Feb 2010. SPE 131787

25. Nobakht, M., Mattar, L., Moghadam, S., Anderson, D.M.: Simplified yet rigorous forecasting of tight/shale gas production in linear flow. In: SPE Western Regional Conference. Society of Petroleum Engineers (SPE), Anaheim, CA, May 2010. SPE 133615

26. Nobakht, M., Mattar, L.: Analyzing production data from unconventional gas reservoirs with linear flow and apparent skin. J. Can. Pet. Technol. 52–59 (2012)

27. Nobakht, M., Clarkson, C.R.: A new analytical method for analyzing production data from shale gas reservoirs exhibiting linear flow: constant pressure production. SPE Reservoir Eval. Eng. J. 370–384 (2012)

28. Cipolla, C.L., Lolon, E.P., Mayerhofer, M.J.: Reservoir modeling and production evaluation in shale-gas reservoirs. In: International Petroleum Technology Conference, Dec 2009, s.n., Doha, Qatar. IPTC 13185-MS

29. Chaudhri, M.M.: Numerical Modeling of multi-fracture horizontal well for uncertainty analysis and history matching: case studies from Oklahoma and Texas shale gas wells. In: SPE Western Regional Meeting. Society of Petroleum Engineers (SPE), Bakersfield, California, Mar 2012. SPE 153888

30. Mayerhofer, M.J., Lolon, E.P., Warpinski, N.R., Cipolla, C.L., Walser, D., Rightmire, C. M.: What is stimulated reservoir volume? Society of Petroleum Engineers (SPE), s.l., Feb 2012. SPE Prod. Oper. J. 25, 89–98. SPE 119890

31. Cipolla, C.L.: Stimulated volume and fracture structure, the keys to shale-gas well performance? In: SPEE 49th Annual Meeting, Amelia Island Plantation, Florida, s.n., June 2011

32. Inamdar, A., Ogundare, T., Purcell, D., Malpani, R., Atwood, K., Brook, K., Erwemi, A.: Pilot Test Stimulation Approach. The American Oil and Gas Reporter, s.l., June 2011
33. Ciezobka, J.: Marcellus Shale Gas Project. RPSEA, Annual Report, s.l., Feb 2012
34. McCulloch, W.S., Pitts, W.: A logical calculus of ideas immanent in nervous activity. Bull. Math. Biophys. **5**, 115–133 (1943)
35. Rosenblatt, F.: The perceptron: probabilistic model for information storage and organization in the brain. Psychol. Rev. **65**, 386–408 (1958)
36. Widrow, B.: Generalization and Information Storage in Networks of Adeline Neurons, Self-Organizing Systems. [book auth.] M.C., Jacobi, G.T., Goldstein, G.D. Yovitz. Self-Organizing Systems. s.n., Chicago, pp 435–461 (1962)
37. Minsky, M.L., Papert, S.A.: Perceptrons. s.n., Cambridge (1969)
38. Hertz, J., Krogh, A., Palmer, R.G.: Introduction to the Theory of Neural Computation. Addison-Wesley Publishing Company, Redwood City, CA (1991)
39. Rumelhart, D.E., McClelland, J.L.: Parallel Distributed processing, Exploration in the Microstructure of Cognition, Vol. 1: Foundations. MIT Press, Cambridge (1986)
40. Stubbs, D.: Neurocomputers. M. D. Comput. **5**(3), 14–24 (1988)
41. Fausett, L.: Fundamentals of Neural Networks, Architectures, Algorithms, and Applications. Prentice Hall, Englewood Cliffs (1994)
42. Barlow, H.B.: Unsupervised learning. Neural Comput. **1**, 295–311 (1988)
43. McCord Nelson, M., Illingworth, W.T.: A Practical Guide to Neural Nets. Addison-Wesley Publishing, Reading, MA (1990)
44. Haykin, S.: Neural Networks and Learning Machines, 3rd edn. Prentice Hall, s.l. (2009)
45. Box, George E.P.: Science and Statistics. American Statistical Association, s.l., Dec 1976. J. Am. Stat. Assoc. **71**(356), 791–799 (1976)
46. Jolliffe, I.T.: Principal Component Analysis, Series: Springer Series in Statistics, 2nd edn. Springer, New York (2002)
47. Shannon, C.E.: A mathematical theory of communication. Bell Syst. Tech. J. **27**, 379–423 (1948)
48. Freeman, E.: The Relevance of Charles Peirce, pp. 157–158. Monist Library of Philosophy, La Sall, IL (1983)
49. Lukasiewicz, J.: Elements of mathematical Logic (Original Title: Elementy logiki matematycznej.]. The MacMillan Company, New York, NY (1963)
50. Black, M.: Vagueness: an exercise in logical analysis. Philos. Sci. **4**, 427–455 (1937)
51. Zadeh, L.A.: Fuzzy sets. Inf. Control **8**, 338–353 (1965)
52. Kosko, B.: Fuzzy Thinking. Hyperion, New York, NY (1991)
53. McNeill, D., Freiberger, P.: Fuzzy Logic. Simon & Schuster, New York, NY (1993)
54. Eberhart, R., Simpson, P., Dobbins, R.: Computational Intelligence PC Tools. Academic Press, Orlando, FL (1996)
55. Ross, T.: Fuzzy Logic With Engineering Applications. McGraw-Hill Inc., New York, NY (1995)
56. Arnsdorf, I.: BloombergBusiness.com. Bloomberg.com. (Online) Bloomberg, 10 8, 2014 (Cited: January 3, 2015). http://www.bloomberg.com/news/2014-10-07/shale-boom-tested-as-sub-90-oil-threatens-u-s-drillers.html
57. Jamshidi, M., et al. (eds.): Fuzzy Logic and Control: Software and Hardware Applications. Prentice Hall, Englewood Cliffs, NJ (1993)
58. Mayr, E.: oward a new Philosophy of Biology: Observations of an Evolutionist. Belknap Press, Cambridge, Massachusetts (1988)
59. Koza, J.R.: Genetic Programming, On the Programming of Computers by Means of Natural Selection. MIT Press, Cambridge, Massachusetts (1992)
60. Fogel, D.B.: Evolutionary Computation, Toward a New Philosophy of Machine Intelligence. IEEE Press, Piscataway, New Jersey (1995)
61. Michalewicz, Z.: Genetic Algorithms + Data Structure = Evolution Programs. Springer, New York (1992). New York

62. Bbauska, R.: Fuzzy and Neural Control. Delft Center for Systems and Control, Delft, Holland (2009)
63. Bezdek, J.: The fuzzy c-mean clustering algorithm. Comput. Geosci. **10**(2–3), 191–203 (1984)
64. Mohaghegh, S., Richardson, M., Ameri, S.: Virtual magnetic resonance imaging logs: generation of synthetic MRI logs from conventional well logs. In: SPE Eastern Regional Conference and Exhibition. Society of Petroleum Engineers (SPE), Pittsburgh, PA, Nov 1998. SPE 51075
65. Mohaghegh, S., Goddard, C., Popa, A., Ameri, S., Bhuiyan, M.: Reservoir characterization through synthetic logs. In: SPE Eastern Regional Conference and Exhibition. Society of Petroleum Engineers (SPE), Morgantown, West Virginia, Oct 2000. SPE 65675
66. Rolon, L.F., Mohaghegh, S.D., Ameri, S. Gaskari, R., McDaniel, B.: Developing synthetic well logs for the Upper Devonian units in Southern Pennsylvania. In: SPE Eastern Regional Conference and Exhibition. Society of Petroleum Engineers (SPE), Morgantown, West Virginia, Sept 2005. SPE 98013
67. Robertson, S.: Generalized Hyperbolic Equation. Society of Petroleum Engineers (SPE), s.l. (1988). SPE 18731
68. Barenblatt, G.I., Zeltov, Y.P., Kochina, I.: Basic concepts in the theory of seepage of homogeneous liquids in fissured rocks. J. Sov. Appl. Math. Mech. **24**, 1286–1303 (1960). 5
69. Root, J.E., Warren, J.P.: The Behavior of Naturally Fractured Reservoirs. Society of Petroleum Engineering, Richardson, Texas, Sept 1963. Soc. Pet. Eng. J. 245–255
70. Kazemi, H.: Pressure transient analysis of naturally fractured reservoirs with uniform fracture distribution. Society of Petroleum Engineers, Richardson, Texas, 1969. SPE J. **9**(4), 451–462 (1969)
71. Rossen, R.H.: Simulation of naturally fractured reservoir with semi-implicit source terms. Society of Petroleum Engineers, Richardson, Texas, June 1977. SPE J. 201–210
72. deSwaan-O, A.: Analytic solutions for determining naturally fractured reservoir properties by well testing. Society of Petroleum Engineers, Richardson, Texas, June 1976. SPE J. 117–122
73. Saidi, A.M.: Simulation of naturally fractured reservoirs. In: Reservoir Simulation Symposium. Society of Petroleum Engineers, San Francisco, CA (1983). SPE 12270
74. Rubin, B.: Accurate simulation of Non-Darcy flow in stimulated fracture shale reservoirs. In: Western Regional Conference. Society of Petroleum Engineers, Anaheim, CA (2010). SPE 132293
75. Cipolla, C.L., Lolon, E.P., Erdle, J.C., Rubin, B.: Reservoir Modeling in Shale-Gas Reservoirs. Society of Petroleum Engineers, Richardson, Texas, 2010. SPE Reservoir Eval. Eng. **13**(4), 638–653 (2010)
76. Ertekin, T., King, G. R., Schwerer, F. C.: Dynamic gas slippage: a unique dual-mechanism approach to the flow of gas in tight formations. Society of Petroleum Engineers, Richardson, Texas, 1986, SPE Formation Eval. **1**(1), 43–52 (1986)
77. Meyer, B.R., Bazan, L.W., Jacot, Lattibeaudiere, M.G.: Optimization of multiple transverse hydraulic fractures in horizontal wellbores. In: SPE Unconventional Gas Conference. Society of Petroleum Engineers (SPE), Pittsburgh, PA, Feb 2010. SPE 131732
78. Cipolla, C.L., Williams, M.J., Weng, X., Mack, M., Maxwell, S.: Hydraulic fracture monitoring to reservoir simulation: maximizing value. In: SPE Annual Technical Conference. Society of Petroleum Engineers (SPE), Florence, Italy, Sept 2010. SPE 133877
79. Samandarli, O., Al-Hamdi, H., Wattenbarger, R.A.: A new method for history matching and forecasting shale gas reservoir production performance with a dual porosity model. In: SPE North American Unconventional Gas Conference. Society of Petroleum Engineers (SPE), The Woodlands, Texas, June 2011. SPE 144335

80. Fisher, M.K., Wright, C.A., Davidson, B.M., Goodwin, A.K., Fielder, E.O., Buckler, W.S., Steinsberger, N.P.: Integrating fracture mapping technologies to optimize stimulations in the Barnett Shale. In: SPE Annual Technical Conference and Exhibition. s.n., San Antonio, Texas (2002). SPE 77441

81. Maxwell, S.C., Urbancic, T.I., Steinsberger, N., Zinno, R.: Microseismic imaging of hydraulic fracture complexity in the Barnett Shale. In: SPE Annual Technical Conference and Exhibition. s.n., San Antonio, Texas (2002). SPE 77440

82. Daniels, J.L., Waters, G.A., Le Calvez, J.H., Bentley, D., Lassek, J.T.: Contacting more of the barnett shale through an integration of real-time microseismic monitoring, petrophysics, and hydraulic fracture design. In: SPE Annual Technical Conference and Exhibition. s.n., Anaheim, California (2007). SPE 110562

83. Kalantari-Dahaghi, A., Esmaili, S., Mohaghegh, S.D.: Fast track analysis of shale numerical models. In: Canadian Unconventional Resources Conference. Society of Petroleum Engineers, Calgary, Alberta, Canada (2012). SPE 162699

84. Mohaghegh, S.D., Abdulla, F.: Production Management Decision Analysis Using AI-Based Proxy Modeling of Reservoir Simulations—A Look-Back Case Study. In: SPE Annual Technical Conference and Exhibition. Society of Petroleum Engineers, Amsterdam, The Netherlands, Oct 2014. SPE 170664

85. Mohaghegh, S.D.: Reservoir simulation and modeling based on artificial intelligence and data mining (AI&DM). J. Nat. Gas Sci. Eng. **3**, 697–705 (2011). 2011

86. Mohaghegh, S.D., Abdulla, F., Abdou, M., Gaskari, R., Maysami, M.: Smart Proxy: an Innovative Reservoir Management Tool; Case Study of a Giant Mature Oilfield in the UAE. In: ADIPEC—Abu Dhabi International Petroleum Exhibition and Conference. s.n., Abu Dhabi, UAE (2015). SPE 177829

87. Shahkarami, A., Mohaghegh, S.D., Hajizadeh, Y.: Assisted history matching using pattern recognition technology. In: SPE Digital Energy Conference and Exhibition. s.n., The Woodlands, Texas, Mar 2015. SPE 173405

88. Shahkarami, A., Mohaghegh, S.D., Gholami, V., Bromhal, G.: Application of Artificial Intelligence and Data Mining Techniques for Fast Track Modeling of Geologic Sequestration of Carbon Dioxide—Case Study of SACROC Unit. In: SPE Digital Energy Conference and Exhibition. s.n., The Woodlands, Texas, Mar 2015. SPE 173406

89. Amini, S., Mohaghegh, S.D., Gaskari, R., Bromhal, G.: Pattern Recognition and Data-Driven Analytics for Fast and Accurate Replication of Complex Numerical Reservoir Models at the Grid Block Level. In: SPE Intelligent Energy Conference and Exhibition. s.n., Utrecht, The Netherlands, Apr 2014. SPE 167897

90. Bazan, L.W., Larkin, S.D., Lattibeaudiere, M.G., Palish, T.T.: Improving Production in Eagle Ford Shale with Fracture Modeling, Increased Conductivity and Optimized Stage and Cluster Spacing Along the Horizontal Wellbore. In: SPE Tight Gas Completions Conference. Society of Petroleum Engineers (SPE), San Antonio, Texas, Nov 2010. SPE 138425

91. Cipolla, C.L., Lolon, E.P., Erdle, J.C., Rubin, B.: Reservoir modleing in shale-gas reservoirs. In: SPE Reservoir Evaluation and Engineering, pp. 638–653. Society of Petroleum Engineers (SPE), s.l., Aug 2010

92. Diaz de Souza, O.C., Sharp, A.J., Martinez, R.C., Foster, R.A., Reeves Simpson, M., Piekenbrock, E.J., Abou-Sayed, I.: Integrated unconventional shale gas modeling: a worked example from the Haynesville Shale, De Soto Parish, North Louisiana. In: Americas Unconventional Resources Conference. Society of Petroleum Engineers (SPE), Pittsburgh, PA, June 2012. SPE 154692

93. Altman, R., Viswanathan, A., Xu, J., Ussoltsev, D., Indriati, S., Grant, D., Pena, A., Loayza, M. and Kirkham, B.: Understanding the impact of channel fracturing in the eagle ford shale through reservoir simulation. In: SPE Latin American and Caribbean Petroleum Engineering Conference. Society of Petroleum Engineers (SPE), Mexico City, Mexico, Apr 2012. SPE 153728

94. Intelligent Solutions, Inc.: (Online) Intelligent Solutions, Inc., 7 Mar 2016. (Cited: March 7, 2016) http://www.intelligentsolutionsinc.com/Technology/TDM.shtml

95. Strickland, R., Purvis, D., Blasingame, T.: Practical aspects of reserves determinations for shale gas. In: North American Unconventional Gas Conference and Exhibition. s.n., The Woodlands, Texas, June 2011. SPE-144357

96. Johnson, P.: Evaluation of wells for re-fracturing treatments. In: Spring Meeting of Southwestern District, Division of production. API Paper 906-5-F. s.n., Dallas, Texas, March 1960

97. Coulter, G.R. Menzie, D.E.: The design of re-frac treatments for restimulation of subsurface formations. In: Rocky Mountain Regional Meeting. Society of Petroleum Engineers—SPE, Casper, Wyoming, May 1973. SPE 4400

98. Mohaghegh, S.D., McVey, D., Aminian, K., Ameri, S. Predicting Well Stimulation Results in a Gas Storage Field in the Absence of Reservoir Data, Using Neural Networks. In: SPE Reservoir Engineering, Vol. Nov, pp. 268–272 Society of Petroleum Engineers (SPE), s.l. (1996)

99. McVey, D., Mohaghegh, S., Aminian, K.: Identification of parameters influencing the response of gas storage wells to hydraulic fracturing with the aid of a neural network. In: SPE Eastern Regional Conference and Exhibition. s.n., Charleston, West Virginia, Nov 1994. SPE 29159

100. Mohaghegh, S. Hefner, H., Ameri, S.: Fracture Optimization eXpert (FOX): How Computational Intelligence Helps the Bottom-Line in Gas Storage. In: SPE Eastern Regional Conference and Exhibition. Society of Petroleum Engineers (SPE), Columbus, Ohio, Oct 1996. SPE 37341

101. Mohaghegh, S., Balan, B., McVey, D., Ameri, S.: A hybrid neuro-genetic approach to hydraulic fracture treatment design and optimization. In: SPE Annual Technical Conference & Exhibition (ATCE). Society of Petroleum Engineers (SPE), Denver, Colorado, Oct 1996. SPE 36602

102. Mohaghegh, S., Platon, V., Ameri, S.: Candidate selection for stimulation of gas storage wells using available data with neural networks and genetic algorithms. In: SPE Eastern Regional Conference and Exhibition. Society of Petroleum Engineers (SPE), Pittsburgh, PA , Nov 1998. SPE 51080

103. Mohaghegh, S., Mohamad, K., Popa, A.S., Ameri, S.: Performance drivers in restimulation of gas storage wells. In: SPE Eastern Regional Conference and Exhibition. s.n., Charleston, West Virginia, Oct 1999. SPE 57453

104. Mohaghegh, S., Gaskari, R., Popa, A., Ameri, S., Wolhart, S.: Identifying best practices in hydraulic fracturing using virtual intelligence techniques. In: SPE Eastern Regional Conference and Exhibition. Society of Petroleum Engineers (SPE), North Canton, Ohio, Oct 2001. SPE 72385

105. Mohaghegh, S., Popa, A., Gaskari, R., Ameri, S., Wolhart, S.: Identifying Successful Practices in Hydraulic Fracturing Using Intelligence Data Mining Tools; Application to the Codell Formation in the DJ Basin. In: SPE Annual Conference and Exhibition (ATCE). Society of Petroleum Engineers (SPE), San Antonio, Texas, Oct 2002. SPE 77597

106. Mohaghegh, S.D.: Essential Components of an Integrated Data Mining Tool for the Oil & Gas Industry, With an Example Application in the DJ Basin. In: SPE Annual Conference and Exhibition (ATCE). Society of Petroleum Engineers (SPE), Denver, Colorado, Oct 2003. SPE 84441

107. Mohaghegh, S.D., Gaskari, R., Popa, A., Salehi. I., Ameri, S.: Analysis of Best Hydraulic Fracturing Practices in the Golden Trend Fields of Oklahoma. In: SPE Annual Conference and Exhibition (ATCE). Society of Petroleum Engineers (SPE), Dallas, Texas, Oct 2005. SPE 95942

108. Mohaghegh, S. D., Gaskari, R.: A Soft Computing-Based Method for the Identification of Best Practices, with Application in Petroleum Industry. In: IEEE International Conference on Computational Intelligence for Measurement Systems& Applications. s.n., Taormina, Sicily, Italy, July 2005

109. Malik, K., Mohagegh, S.D., Gaskari, R:. An Intelligent Portfolio Management Approach to Gas Storage Field Deliverability Maintenance and Enhancement; Part One Database Development & Model Building. In: SPE Eastern Regional Conference & Exhibition. Society of Petroleum Engineers (SPE), Canton, Ohio, Oct 2006. SPE 104571

110. Reeves, S.R., Hill, D.G., Tiner, R.L., Bastian, P.A., Conway, M.W., Mohaghegh, S.D.: Restimulation of Tight Gas Sand Wells in the Rocky Mountain Region. In: SPE Rocky Mountain Region Meeting. Society of Petroleum Engineers (SPE), Gillette, Wyoming, May 1999. SPE 55627

111. Reeves, S.R., Hill, D.G., Hopkins, C.W., Conway, M.W., Tiner, R.L., Mohaghegh, S.D.: Restimulation Technology for Tight Gas Sand Wells. In: SPE Technical Conference and Exhibition (ATCE). Society of Petroleum Engineers (SPE), Houston, Texas, Oct 1999. SPE 56482.

112. Mohaghegh, S., Reeves, S., Hill, D.: Development of an Intelligent Systems Approach to Restimulation Candidate Selection. In: SPE Gas Technology Symposium. Society of Petroleum Engineers (SPE), Calgary, Alberta, Apr 2000. SPE 59767

113. Reeves, S., Bastian, P., Spivey, J., Flumerfelt, R., Mohaghegh, S., Koperna, G.: Benchmarking of Restimulation Candidate Selection Techniques in Layered, Tight Gas Sand Formations Using Reservoir Simulation. In: SPE Annual Technical Conference and Exhibition (ATCE). Society of Petroleum Engineers (SPE), Dallas, TX, Oct 2000. SPE 63096

114. Jacobs, T.: Halliburton reveals refracturing strategy. J. Pet. Technol. (JPT) 40–41 (2015) (November)

115. Siebrits, E., et. al.: Refracture Reorientation Enhances Gas Production in Barnett Shale Tight Gas Wells. In: SPE Annual Technical Conference and Exhibition (ATCE). Society of Petroleum Engineers (SPE), Dallas, Texas, Oct 2000. SPE 63030

116. Bell, G., Hey, T., Szalary, : Beyond the data deluge. Science 23, 1297–1298 (2009)

117. Pelham Box, G.E.: Science and Statistics, p. 792 (1976)

118. Thakur, G.C.: What Is Reservoir Management? Society of Petroleum Engineerd (SPE), Richardson, Texas. J. Pet. Technol. 48(6), 520–525 (1996)

119. Chevron Corporation: Reservoir Management. Chevron Corporation Website. (Online) Chevron Corporation Website. http://www.chevron.com/deliveringenergy/oil/reservoirmanagement/ (2012)

120. Mata, D., Gaskari, R., Mohaghegh, S.D.: Field-Wide Reservoir Characterization Based on a New Technique of Production Data Analysis, Lexington, Kentucky, Oct 2007. SPE 111205

121. Gomez, Y., Khazaeni, Y., Mohaghegh, S.D., Gaskari, R.: Top-Down Intelligent Reservoir Modeling (TDIRM), New Orleans, Louisiana (2009). SPE 124204

122. Gaskari, R., Mohaghegh, S.D., and Jalali, J.: An Integrated Technique for Production Data Analysis (PDA) with Application to Mature Fields. Society of Petroleum Engineers (SPE), Richrdson, Texas, Nov 2007, SPE Prod. Oper. J. 22(4), 403–416

123. Mohaghegh, S.D., Gaskari, R.: An intelligent system's approach for revitalization of brown fields using only production rate data. Int. J. Eng. 22, 89–106 (2009)

124. Kalantari, A.M., Mohaghegh, S.D., Khazaeni, Y.: New Insight into Integrated Reservoir Management using Top-Down, Intelligent Reservoir Modeling Technique; Application to a Giant and Complex Oil Field in the Middle East. In: SPE Western Regional Conference & Exhibition, Anaheim, California, May 2010. SPE 132621

125. Khazaeni, Y., Mohaghegh, S.D.: Intelligent Production Modeling Using Full Field Pattern Recognition. Society of Petroleum Engineers (SPE), Richardson, Texas, Dec 2011. SPE Reser. Eval. Eng. J. 14(6), 735–749

126. Maysami, M., Gaskari, R., Mohaghegh, S.D.: Data Driven Analytics in Powder River Basin, WY. In: SPE Annual Technical Conference and Exhibition, New Orleans, Louisiana (2013). SPE 166111

127. Zargari, S., Mohaghegh, S.D.: Development Strategies for Bakken Shale Formation. In: SPE Eastern Regional Conference & Exhibition, Morgantown. s.n., Morgantown, West Virginia (2010). SPE 139032

128. Esmaili, S., Kalantari, M., Mohaghegh, S.: Modeling and History Matching Hydrocarbon Production from Marcellus Shale using Data Mining and Pattern Recognition Technologies. In: SPE Eastern Regional Conference, Lexington, Kentucky (2012). SPE 161184

129. Grujic, O., Mohaghegh, S.D., Bromhal, G.: Fast Track Reservoir Modeling of Shale Formations in the Appalachian Basin. In: SPE Eastern Regional Conference & Exhibition, Application to Lower Huron Shale in Eastern Kentucky, Morgantown, West Virginia (2010). SPE 139101

130. Kalantari, A.M., Mohaghegh, S.D.: A new practical approach in modeling and simulation of shale gas reservoirs: application to New Albany Shale. Int. J. Oil Gas Coal Technol. 4, 104–133 (2011). 2

131. Mohaghegh, S.D., Grujic, O., Zargari, S., Kalantari, A.M., Bromhal, G.: Top-down, intelligent reservoir modeling of oil and gas producing shale reservoirs; case studies. Int. J. Oil Gas Coal Technol. 5(1), 3–28

132. Haghighat, A., Mohaghegh, S.D., Gholami, V., Moreno, D.: Production Analysis of a Niobrara Field Using Intelligent Top-Down Modeling. In: SPE Western North American and Rocky Mountain Joint Regional Meeting, Denver, Colorado (2014). SPE 169573

133. Aurenhammer, F.: Voronoi diagrams—a survey of a fundamental geometric data structure. ACM Comput. Surv. 23(3), 345–405 (1991)

134. Sayarpour, M., Kabir, C.S., Lake, L.W.: Field Applications of Capacitance-Resistance Models in Waterfloods. Society of Petroleum Engineers, s.l., Dec 2009, SPE J. (2009)

135. Wooldridge, M.J., Jennings, N., Müller, J.P.: Stan Franklin, Art Graesser. Is it an Agent, or just a Program? A Taxonomy for Autonomous Agents. Intelligent agents III agent theories, architectures, and languages, pp. 21–35. Springer, Berlin (1997)

136. Stufflebeam, R., Mills, F.: Introduction to Intelligent Agents. Consortium on Cognitive Science Instruction. (Online) The Mind Project (Cited: June 23, 2015). http://www.mind. ilstu.edu/curriculum/ants_nasa/intelligent_agents.php

137. Swan, A.R.H., Sandilands, M.: Introduction to Geological Data Analysis. Blackwell, Oxford (1995)

138. Jensen, J.L., Lake, L.W., Corbett, P.W.M., Goggin, D.J.: Statistics for Petroleum Engineers and Geoscientists, 2nd edn. Elsevier, Amsterdam (2000)

139. Davis, J.C.: Statistics and Data Analysis in Geology, 3rd edn. Wiley, New York (2002)

140. King M.J., Datta-Gupta, A.: Streamline Simulation: Theory and Practice. Society of Petroleum Engineers, Richardson, Texas (2007)

141. Jalali, J., Mohaghegh, S.D., Gaskari, R.: Identifying Infill Locations and Underperformer Wells in Mature Fields using Monthly Production Rate Data, Carthage Field, Cotton Valley Formation, Texas. Society of Petroleum Engineers, Canton, Ohio (2006). SPE 104550

142. Gaskari, R., Mohaghegh, S.D.: Estimating major and Minor Natural Fracture pattern in Gas Shales Using Production Data. Society of Petroleum Engineers, Canton, Ohio (2006). SPE 104554

143. Mohaghegh, S.D.: Top-Down Intelligent Reservoir Modeling (TDIRM); A New Approach in Reservoir Modeling by Integrating Classis Reservoir Engineering with Artificial Intelligence and Data Mining Techniques. In: American Association of Petroleum Geologists (AAPG), Denver, Colorado (2009)

144. Mohaghegh, S.D., Bromhal, G.: Top-Down Modeling; Practical, Fast Track, Reservoir Simulation & Modeling for Shale Formations. In: AAPG/SEG/SPE/SPWLA Hedberg Conference, Austin, Texas (2010)

145. Esamili, S., Mohaghegh, S.D.: Full field reservoir modeling of shale assets using advanced data-driven analytics. Geosci. Front. **1**, 11 (2015) (Elsevier, s.l.)
146. Mohaghegh, S.D., Gaskari, R., Maysami, M., Khazaeni, Y.: Data-Driven Reservoir Management of a Giant Mature Oilfield in the Middle East. Society of Petroleum Engineers, Amsterdam, Holland (2014). SPE 170660
147. Al-Sharhan, A.S.: Bu Hasa Field—United Arab Emirates, Rub al Khali Basin, Abu Dhabi. [book auth.] N.H. Foster and Beaumont. Treaties of Petroleum Geology, Atlas of Oil and Gas Fields. AAPG, s.l. (1993)
148. Dickerson, M.T., Goodrich, M.T., Dickerson, M.D., Zhuo, Y.D.: Round-trip voronoi diagrams and doubling density in geographic networks. Trans. Comput. Sci. **14**, 211–238 (2011)
149. Höppner, F., et al.: Fuzzy Cluster Analysis: Methods for Classification, Data Analysis and Image Recognition. Wiley IBM PC Series, s.l. John Wiley & Sons (1997)
150. Shannon, C.E.: A Mathematical Theory of Communication. Bell Syst. Tech. J. **27**, 379–423, 623–656 (1948)
151. Mohaghegh, Shahab D.: Reservoir modeling of shale formations. J. Nat. Gas Sci. Eng. **12**, 22–33 (2013)

Printed in the United States
By Bookmasters